WITHDRAWN

Techniques and Applications of Digital Watermarking and Content Protection

For quite a long time, computer security was a rather narrow field of study that was populated mainly by theoretical computer scientists, electrical engineers, and applied mathematicians. With the proliferation of open systems in general, and of the Internet and the World Wide Web (WWW) in particular, this situation has changed fundamentally. Today, computer and network practitioners are equally interested in computer security, since they require technologies and solutions that can be used to secure applications related to electronic commerce. Against this background, the field of computer security has become very broad and includes many topics of interest. The aim of this series is to publish state-of-the-art, high standard technical books on topics related to computer security. Further information about the series can be found on the WWW at the following URL:

http://www.esecurity.ch/serieseditor.html

Also, if you'd like to contribute to the series by writing a book about a topic related to computer security, feel free to contact either the Commissioning Editor or the Series Editor at Artech House.

For a listing of recent titles in the *Artech House Computer Security Series,* turn to the back of this book.

Techniques and Applications of Digital Watermarking and Content Protection

Michael Arnold
Martin Schmucker
Stephen D. Wolthusen

Artech House
Boston • London
www.artechhouse.com

Library of Congress Cataloging-in-Publication Data
Arnold, Michael (Michael Konrad), 1964–
    Techniques and applications of digital watermarking and content protection / Michael
Arnold, Martin Schmucker, Stephen D. Wolthusen.
        p. cm.—(Artech House computer security series)
    Includes bibliographical references and index.
    ISBN 1-58053-111-3 (alk. paper)
        1. Computer security. 2. Digital watermarking. 3. Data protection. I. Schmucker,
Martin. II. Wolthusen, Stephen D. III. Title.

QA76.9.A25A76 2003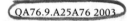
    005.8—dc21   2003049577

British Library Cataloguing in Publication Data
Arnold, Michael
    Techniques and applications of digital watermarking and content protection.—
    (Artech House computer security series)
    1. Digital watermarking
    I. Title II. Schmucker, Martin III. Wolthusen, Stephen D.
    005.8

    ISBN 1-58053-111-3

Cover design by Yekaterina Ratner

© 2003 ARTECH HOUSE, INC.
685 Canton Street
Norwood, MA 02062

International Standard Book Number: 1-58053-111-3
Library of Congress Catalog Card Number: 2003049577

10 9 8 7 6 5 4 3 2 1

To our parents

Contents

6 | Digital watermarking for other media 115

7 | Attacks and benchmarks of digital watermarking systems 147

Preface

Few research subjects in computer science are as inextricably linked with problems from other disciplines, ranging from economic theory to intellectual property law and even politics, as the study of protection mechanisms for intellectual property.

The present book therefore not only introduces a number of techniques for such mechanisms (primarily focused on digital watermarking but also including other mechanisms). Rather, it strives to place these techniques into a context provided by application scenarios and their requirements, as well as the legal and—to a modest extent—ethical contexts all such research and development must be conducted in. Given this premise, particular attention is paid to clearly showing the limits of the currently known techniques and algorithms within the particular context of intellectual property protection.

The book itself is structured in such a way that it can be read both in sequence and in a topical fashion. Chapter 1 discusses the legal background and historical origins of intellectual property and forms the basis for the remainder of the book, which consists of two logical segments. The first and larger segment is dedicated to digital watermarking. Here, Chapter 2 provides an introduction to general concepts while Chapter 3 describes application scenarios for digital watermarking.

Following this general introduction, three separate chapters introduce media-specific techniques: Chapter 4 for still images, Chapter 5 for audio data, and finally Chapter 6 for several other media types, including video and formatted text, but with an emphasis on three-dimensional data sets. Based on this exposition, Chapter 7 concludes the discussion of digital watermarking with a detailed analysis of attack mechanisms and benchmarking approaches.

The second segment provides an introduction to various other protection techniques. Chapter 8 provides a discussion of copy protection schemes in the case of software in general purpose computers, as well as techniques for analog media. The remainder of Chapter 8 discusses protection mechanisms for digital representations of multimedia data, with particular emphasis on audio data.

Chapter 9 then introduces the concepts and application requirements for digital rights management systems and discusses several examples for this purpose. Given the fundamental requirements of digital rights management systems, particular attention is paid to the issue of tamper resistance and particularly the limits thereof in commercially viable systems.

Finally, Chapter 10 provides an assessment of the technologies presented thus far, as well as an outlook of the research challenges yet ahead.

The authors hope that the present book can serve as both an introduction to the field to both students and researchers, providing sufficient background for further studies and guidance for the assessment and—if necessary—selection of technical means for intellectual property protection for decision makers and the general public.

Acknowledgments

The authors would like to thank Dr. J. L. Encarnação, who established the security technology department at the Fraunhofer Institute for Computer Graphics, as well as the current and former department heads Dr. C. Busch and Dr. E. Koch for providing a stimulating research environment. Also, the authors would like to acknowledge the contributions, direct or indirect, of present and former researchers at Fraunhofer-IGD in the area of digital watermarking, specifically O. Benedens, S. Burgett, W. Funk, N. Schmitt, and J. Zhao (who provided the suggestion to write this book).

P. Neugebauer and K. Klein assisted with the reconstruction of three-dimensional geometry data from physical models. S. Krusche, S. Lang, and O. Lobisch created several drawings and figures. V. Ochs, S. Rogge, E. Singer, and A. Zeidler provided valuable aid in proofreading and checking the consistency of bibliographic information.

The editorial and acquisitions staff at Artech House shepherded this book into what is hoped to be a useful form. The authors would particularly like to acknowledge the acquisitions editor T. Pitts and the assistant editor T. Ruonamaa, whose advice and management helped considerably in bringing this book to fruition.

Acknowledgments

CHAPTER

1

Contents

Introduction

While a considerable portion of economic resources is dedicated to the creation of intellectual property, particularly in industrial societies, the cost of reproducing such intellectual creations (i.e., the marginal cost of production) typically constitutes only a small fraction of the creation. As such, these creations share some properties with public goods in that they are—at least without additional means—nonexcludable and nonrival in consumption [1]. However, a creator may expect some measure of compensation for the resources expended in the process creation, since otherwise the incentives for new creations, at least based on economic motives, are severely curtailed. This compensation for various forms of intellectual property has been accorded through various, mainly legal techniques since the middle ages; a selection of intellectual property rights types is shown in Table 1.1.

The oldest type of protection is presumably the trade or craft secret; the government of Renaissance Venice sought to place limits on the migration of craftsmen skilled in glassmaking from Venice as well as on these craftsmen taking on apprentices from outside of Venice. This attempt at secrecy confined to a location and guild gradually and partly gave way to the granting of patents beginning in the thirteenth century [2], although confidential corporate information for which other means of protection are either not desired or feasible to this day must be protected as such; the legal protection of trade secrets is relatively weak compared to that afforded to other types of intellectual property [3].

Other areas that are protected legally by special regulations include digital databases (this is limited to Europe as of 2002 due

Table 1.1 Intellectual Property Right Types

Subject	Type
Confidential corporate information	Trade secrets
Original works of authorship	Literary and artistic property
Inventions for industrial application	Industrial property
New biological varieties	Sui generis protection
Digital databases	Sui generis protection
Semiconductor layout	Sui generis protection

From [1].

to the European Council Directive 96/9/EC) and the topography of semiconductors [4], as well as plant breeders for newly bred distinguishable plant varieties.

As noted above, another major mechanism for the protection of intellectual property is that of patents, trademarks, and designs that are granted by most states as a legal title for making exclusive commercial use of a creation for a limited time in exchange for the publication of the subject of the patent in such a way that it can be reproduced. The granting of (modern) patents is typically tied to a provision that the patent be nonobvious to someone versed in the same field of work, new, and commercially applicable. This barrier is less stringent for utility models or industrial designs where the criteria for innovation are lessened.

However, the type of intellectual creation this book is mainly concerned with is literary and artistic property. This type is legally protected in most jurisdictions by copyrights and neighboring rights and has been the subject of international treaties harmonizing the rights afforded to the creators or rights owners since the seminal Berne Convention for the Protection of Literary and Artistic Works was signed in 1886 [5].

The term itself contains attributes that are not easily quantified, since they contain a value judgment—one of the reasons for the European Community's decision to protect digital databases was that databases were judged to be neither literary nor artistic artifacts—and this judgment may differ between jurisdictions. The type of expression is a priori not limited; while legal protection was first afforded to literary works (see Section 1.1), the protection encompasses all types of artistic expression, such as novels, short stories, poems, plays, music, motion pictures, software, drawings, paintings, sculptures, and architectural work or geometric models thereof.

1.1   The origins of copyright protection

The Psalter of St. Columba—also known as the Cathach or the Battle Book of the Clan O'Donnell—is the oldest extant Irish manuscript of the Psalter and at the same time may also be considered the first example of copyright infringement known in the English-speaking world (the manuscript can be dated to between A.D. 560 and 630; traditionally the date of 567 is given). Tradition has Columba of Iona copying the Psalter of abbot Finnian without permission. When St. Finnian learned about this he demanded that the copy be surrendered to him, but Columba refused. As a result, Finnian appealed to High King Diarmait Mac Cerbhaill. The king gave the judgment, "To every cow belongs her calf, therefore to every book belongs its copy" and ordered Columba to deliver the copy to Finnian, but Columba did not comply until the battle of Culdreimhe had been fought over the matter[1] [6].

The United States Constitution states that "The Congress shall have Power . . . To promote the Progress of Science and useful Arts, by securing for limited Times to Authors and Inventors the exclusive Right to their respective Writings and Discoveries." The origin of this concept—but not of the noble sentiment of promoting progress in the arts and sciences—in the Anglo-American legal system (similar restrictions also existed in France) stems from a royal charter granted by Mary Tudor, Queen of England, to the Stationer's Company [7] in 1557. This charter limited the right to print books to the members of the company.[2] The intent behind this privilege was primarily to exert censorship; the commercial interests of the publishers were of secondary interest only. Even after the repealing of the 1662 Licensing Act in 1681, the Stationer's Company retained control over the printing trade through the use of a bylaw establishing rights of ownership for books registered to its members. This common law mechanism was supplanted in 1710 by the Statute of Anne enacted in 1709. The Act of Parliament granted authors copyright over their work initially for 14 years and was the first copyright legislation in the current sense; in most European states the rights of the authors were recognized only partially until the French Revolution [8].

---

1. The surviving leaves of the manuscript and its shrine are now in the custody of the Royal Irish Academy in Dublin.
2. The charter states "Euery boke or thinge to be allowed by the stationers before yt be prynted," that is, books had to be entered into the stationer's register before they could be printed. At least one publisher, William Carter, was executed for publishing a book illegally.

Modern copyright regulations, on the other hand, emphasize the incentives for creators of works provided by the potential for remuneration from exploitation of the rights to the work as their rationale, while conversely society stands to benefit from the creation of new works. Copyright, as defined in the tradition of the Berne Convention,[3] gives the creator of a work the following exclusive, intangible rights:

- *Reproduction right:* The right to create copies of a protected work;

- *Derivative right:* The right to create new works, either derivative works or adaptations of the work, based on a protected work;

- *Distribution right:* The right to sell or otherwise distribute copies to the public;

- *Performance and display rights:* The right to perform a protected work, such as a musical composition, or to display a protected work, such as a photograph, in public. These rights vary between jurisdictions and can also depend on the type of work.

While the rights initially belong to the creator, they can—as with any other property right—be transferred, sold, or otherwise exploited for economic benefit either in part or in total, although the continental European (primarily French) tradition regards some rights (*droits d'auteur*) as nontransferable.

Unlike other intellectual property rights, copyright is conferred automatically without the need to formally register a work among the signatories of the Berne Convention since the Berlin Act of 1908 and its successors culminating in the World Intellectual Property Organization (WIPO) Copyright Treaty [9], although at some time jurisdictions, including the United States, have required the owners of rights to a work to affix a copyright notice to each copy of the work to ensure that their rights were maintained.

The privileges granted by copyrights are, however, fettered in most jurisdictions by a "fair use" doctrine (e.g., established in the United States by *Sony Corp. of Am. v. Universal City Studios, Inc.*, 464 U.S. 417 (1984) in which the Supreme Court stated that "the fair use doctrine exists because copyright law extends limited proprietary rights to copyright owners only to the extent necessary to ensure dissemination to the public" with regard to using video cassette recorders for personal use of broadcast material). This creates

---

3. The U.S. was not a party to the 1886 treaty and remained governed by the 1790 Copyright Act, resulting in a lack of protection of European works in the United States and vice versa until separate bilateral agreements were reached.

exceptions to the exclusive rights held by the rights owners, such as reproduction for research and scholarship, including comment and criticism or news reports.

1.2   The protection of intellectual property through technical means

While legal means for the protection of intellectual property rights have been largely effective—at least with regard to reproduction and derivation in the domain for which they were initially created, that is, published material such as books—a number of issues make the sole reliance on legal means at least partially questionable [10].

The photomechanical reproduction of entire books on a small scale such as for personal use (as opposed to, for example, large-scale repeated reproduction as may be found in educational settings), even with high quality xerography or its predecessor techniques such as hectography, required a considerable effort and was therefore economical only where labor cost did not enter into consideration. Due to the quality degradation inherent in the process (e.g., by geometric distortions or optical imperfections), the result of such a reproduction was also noticeably less attractive than its original. Consequently, the monetary loss for the rights owners was limited in such situations. The potential losses are significantly larger in the case of unauthorized ("pirated") editions (including translations into foreign languages) given the number of copies distributed; however, not only is the investment to be made by the pirate publisher larger, the distribution and sale of the pirated editions is likely to come to the attention of the rights owner, who can then (assuming the copyright violation occurs in a jurisdiction with suitable statutes) initiate legal proceedings.

Similar observations can be made with regard to other analog physical media that are subject to copyright protection such as audio material; the extent of the violations of rights owners' duplication rights were severely limited by the inconvenience of operating reel-to-reel tape recorders, while personal copies were likely to fall under the fair use doctrine, as was the case with broadcast video material recorded for personal use on video cassette recorders (see Section 1.1).

In the case of analog audio material, the advent of compact cassette recorders[4] markedly lowered the technical barrier for duplication of compact

---

4. Philips first exhibited the compact cassette at the 1963 Berlin Radio Exhibition (IFA); while Grundig and Telefunken also showed their new DC cassette system at the same show, the compact cassette rapidly gained acceptance among other manufacturers and customers. Originally intended for recording monaural voice, it was later enhanced to reproduce stereo audio and further gained in quality with the adoption of Dolby Laboratories noise reduction systems.

cassettes or transfer from other media; both individuals and large-scale pirate operations could conveniently create copies with readily available devices and minimal investments. Despite their obvious inferiority in quality to newer media formats, compact cassettes still dominated the pirate sales of audio material by a considerable margin (65%) as late as 2001 [11].

The observation that convenience and cost figure prominently in both small-scale individual acts of piracy and in organized piracy can again be made with regard to the compact disc (CD), first released in 1982.[5] Small-scale CD production (and hence piracy) only became possible in 1988, albeit at considerable cost.[6] Since then, standalone CD recording devices have become a consumer product, high-speed CD recording functionality has become a standard feature of computer-attached optical media devices, and the cost of blank media has declined precipitously. With the removal of the barriers previously in place with analog recording equipment, that is, primarily the degradation in quality of the copied material particularly after several generations of analog copying, as well as the speed at which a medium can be duplicated (entire CDs can be copied unattended in less than 5 minutes), the threat to proper remuneration of rights owners is considerable.

In terms of the number of units of pirated CDs, the recordable CD (CD-R) does not, however, represent the biggest challenge. Instead, this threat emanates from CDs pressed from regular glass masters in CD production plants worldwide (the International Federation of the Phonographic Industry (IFPI) estimates worldwide sales in 2001 of pressed pirate CDs at 475 million, CD-R at 165 million units [11]) in the form of organized piracy. This represents a severe challenge, as the production equipment may not always reside in jurisdictions where copyright violations are prosecuted vigorously. Perpetrators will copy not only the digital content but also the physical representation (e.g., CD covers), resulting in—at least at first glance—a faithful copy of the original material, unlike the easily recognizable CD-Rs. This allows the pirates to supply unsuspecting consumers with illegal copies for which the consumer will pay retail prices, resulting in verifiable losses to the rights owners that cannot necessarily be assumed in the case of CD-R copies.

The availability of inexpensive high-bandwidth Internet connectivity and processing power to end users has resulted in the ability to display or

---

5. Core patents for the compact disc are held by Philips and Sony; the specification ("Red Book," CEI IEC 908) describing the physical properties of the medium and digital audio encoding dates from 1980.
6. The Yamaha Programmable Disc Subsystem formed the basis for most early devices and cost approximately $35,000 in 1988, not including the supporting facilities for mastering; the cost of a blank medium was approximately $100.

otherwise process, transmit, and store all relevant media formats, particularly audio and video material using readily available general purpose systems [12]. Fast computers permit the use of lossy compression algorithms which reduce the bandwidth and storage capacities required to levels within the capacity of most individuals.

In the case of software this problem has been a constant presence since the advent of standardized computer systems; even before the spread of bulletin board systems (when paper (or mylar) tape was still the dominant medium), the ratio of copies sold to copies distributed of certain software packages was a cause for concern [13]. Particularly in the area of small computers, this resulted in an ongoing competition between the creators of copy protection mechanisms and their adversaries [14] (see Chapter 8).

Initially confined in the area of multimedia data to audio data after the popularization of the Motion Picture Expert Group (MPEG) audio compression standard [15], the phenomenon has since been extended to video data with the advent of the MPEG-4 standard [16]. Given that the motion picture industry relies heavily on a sequence of releases in different geographical markets and on the sequential use of distribution forms, including movie theaters, rental, and eventually sale to individuals, the availability of pirated copies with a quality apparently accepted by consumers as equivalent to screening in movie theaters, and hence several months in advance of the scheduled sales to individuals, is therefore particularly damaging.

Another effect of the use of networked computer systems for the dissemination of copyrighted material is that the works can be exchanged among individuals with great ease, in many cases automatically. As a result, the distribution of pirated works has become a simple and automated process that is difficult to track across national borders and in which individuals offering or receiving pirated material are difficult to identify; the twin factors of convenience and interoperability that were instrumental in the rise in popularity of other mechanisms are present in such network-based exchange mechanisms.

The observations listed above can lead to the conclusion that technical means for the protection of intellectual property are called for, since organizational means may either not be available (in the case of the intellectual property that is distributed for sale) or insufficient.

Such technical mechanisms can be classified in a taxonomy as follows:

1. *Copy protection:* This is the most direct form of control exertion. An entity is sold or licensed a fixed number of copies; the mechanism must ensure that no additional replication takes place. Several subtaxa can be identified:

> ▸ *Analog physical media:* The medium must permit the faithful reproduction of the work on designated devices and yet modify this signal in such a way that it cannot be reproduced.

> ▸ *Analog ephemeral data:* Broadcast signals must be transmitted in such a way that only designated devices can reproduce the work.

> ▸ *Digital physical media:* Copying the medium must require additional operations in excess of those required for simple reproduction of the work while requiring the presence of the additional information or features for reproduction.

> ▸ *Digital ephemeral data:* The data required for reproducing the work must be processed by a device that does not permit replication of protected works.

2. *Usage monitoring:* Each individual instance of an operation or set of operations defined as a usage must be recorded or communicated in such a way that the information can subsequently be used by the rights owner of a work or the owner's agent.

3. *Distribution tracing:* The creation of a copy and subsequent transmission thereof to another device or individual or the forwarding of the original instance of the work must result in the creation of information recording a feature identifying the source, and may also result in the creation of information recording a feature identifying the destination of the transmission.

4. *Usage control:* Each individual instance of an operation or set of operations defined as a usage must be subject to the approval of the rights owner of a work or an agent thereof.

The mechanisms that can be employed to reach these objectives range from physical features of media or devices that pose difficulties in copying, to elaborate digital rights management schemes tracing the distribution of individual pirated digital copies of works using fingerprinting and digital watermarking techniques.

While the precise definitions of each entry in the taxonomy may be subject to contention, the potential contradictions inherent in most of the requirements for protection mechanisms should be apparent.

Copy protection for digital representations requires that the device about to perform the copying cooperates in the blocking of disallowed copying.

Even when relying on physical characteristics of media such as tolerances for nonstandard recording or the use of other mechanisms that deviate from the standards established for media and data formats, there exists an implicit reliance on specific device features (albeit post factum in the latter case). Not only is this undesirable, since it may exclude potential customers or result in customer complaints and litigation, such schemes also rely on the integrity of the cooperating devices.

The integrity is threatened, since in many scenarios (e.g., sale of movies to the general public) the device is under the physical control of a potential adversary for an unlimited time. While tamper resistance can be achieved to some extent even in consumer devices, the cost of such measures limits the strength of mechanism that can be obtained. Given that a major source of losses for publishers of audio and video material is organized crime, which can be assumed to have adequate financial resources, this limits the reliance that can be placed on tamper resistance [17]. Moreover, even a device that performs a successful self-destruction or disabling sequence on the detection of an attack is not adequate, since the adversary has, in principle, access to unlimited numbers of identical devices for further probing and development of attacks (see Section 9.2 for a further discussion of this problem).

While the majority of this book consists of enabling technologies for achieving the objectves given in the taxonomy above (albeit also touching upon related uses for said technologies), it is only in the awareness of this general limitation found in the above-mentioned scenarios that one can legitimately consider technical mechanisms to achieve the goals of content protection. Particularly in the case of digital ephemeral data, the protection afforded by a mechanism that would thwart an individual nontechnical user in terms of the expertise required to circumvent or tamper with the protection mechanism is not necessarily sufficient to restrain even this user, since if even a single individual devises such a circumvention scheme and makes it available to others, the barrier for nontechnical users becomes the ability to locate sources for circumvention schemes that can be applied with minimal skills. Therefore, the statement that protection schemes are intended to "keep honest people honest," typically stated in acknowledgment of the infeasibility of sufficient protection against qualified adversaries, becomes dubious at best.

Usage monitoring, mainly of interest in the area of multimedia data where monitoring solutions (e.g., for broadcasting) are well established for the determination of royalty payments based on playlists and the verification of broadcasting of commercials in accordance with contracts, is problematic in that most ways of formatting and attaching metadata on the media themselves require standardization of exchange formats, do not survive encoding transitions (particularly to the analog domain), and are easily removed. Digital watermarking (i.e., the embedding of a payload signal within the carrier signal itself) and fingerprinting (i.e., the derivation of characteristic patterns for identification of media data from the signal itself) represent solutions for this application scenario that are independent of media formats and encodings and can be implemented unobtrusively. Digital watermarks also can assist in the process of distribution tracing if the markings in the media identify the path in the distribution graph.

Usage control implies that the control mechanism overrides operations counteracting the interests of the user of the system providing the controlled access and usage facility. It is thus a highly intrusive mechanism, as it may not only counteract other property rights—such as preventing an individual from reproducing works that are created or owned by that individual—but because it has, as in the case of usage monitoring mechanisms, the potential to violate privacy rights and expectations in an egregious manner.

Both research and development activities in the field of content protection have a moral, if not legal, obligation to be aware of the implications inherent in the techniques they are investigating and ultimately deploying and of the responsibilities, direct or indirect, that are associated with the potential for misuse.

Protection mechanisms may result in the works protected being dependent on specific devices or software (which in turn depends on specific devices and ancillary software) that are likely to change rapidly in response to advances in technology and the need to counteract the successful circumvention of existing protection mechanisms on the part of adversaries. Such changes will, over the course of time, very likely result in devices that are unable to reproduce works recorded or stored for use with earlier devices and protection mechanisms. This is problematic in that customers who bought copies or licenses to works may be precluded from migrating these works from one media format to another and thus their property rights are violated. A larger problem, however, lies in the danger to long-term archiving of works even after the rights owners have lost the expectation of financial gain. The time spans covered by copyright law in most jurisdictions cover a multitude of generations in both the devices for reproduction of works and the protection mechanisms; additionally, even after the expiry of a copyright claim, it may not be legal or feasible to reverse-engineer or otherwise circumvent a protection mechanism.

The very existence of usage tracing and control mechanisms also implies risks of misappropriation. The information on which work is used by a certain individual is of interest for commercial entities but can easily violate the expectation of privacy individuals have in the confines of their homes; while in many similar circumstances (e.g., in the use of electronic payment systems) this lack of privacy is accepted to some extent, the level of detail of intrusion and the inability to evade the privacy-depriving mechanism (as in the case of cash transactions for payment) are cause for concern.

Lastly, it is also important to note the potential that particularly usage tracing and control mechanisms have for misuse by political entities. Outright censorship or the mere knowledge that any access to information is subject to surveillance by governmental entities can result in a severe curtailment of individual freedoms of expression and ultimately thought due to a lack or selective availability of relevant information. The origins of copyright at the nexus of censorship and the protection of rights owners remains a blemish on any technology or mechanism.

1.3   Integrity and authenticity

While the primary focus in content protection is on aspects of copyright protection in the literal sense, there are other threats to intellectual property holders that can have direct or indirect detrimental effects.

The integrity of data, or specifically that of protected works, is of concern in many situations. In the case of creations with artistic character, the rights owner may wish to ensure that the work or duplicate of a work remains true to the intentions of the rights owner and that manipulation is either prevented or discoverable.

However, the issue of integrity also arises in other situations where it cannot be ensured by procedural means, such as restricting access to authorized personnel and securing the chain of custody. This may be the case when a signal is recorded by a sensor that becomes relevant as evidence at a later time after being distributed to other parties without means for verifying the origin or pedigree of the recorded signal.

One area in which the manipulation of signals has a long history is that of photography and video manipulation. Frequently relied upon as prima facie evidence in many situations (such as in news media [18–21], purchasing decisions [22], as positive or negative criminal evidence, or in a political context), a well-executed manipulation is deemed as credible by most individuals without specific training as photo or intelligence analysts [23–26].

The issue of integrity of a work is also closely interrelated to that of authenticity. The origin of a work and its authorship can be documented using many of the same techniques used for the protection of integrity, but there exist scenarios in which this is insufficient. One such scenario, familiar from mail fraud in the physical domain, is the fraudulent use of certain designs and appearances to evoke the impression of governmental authority or a commercial entity. Such a technique was used successfully by an unidentified entity to raise the stock price of PairGen by 31% on April 7, 2000; the means through which this was perpetrated was a node on the Internet (referred to only by a numerical address) to which a fraudulent news item about the acquisition of PairGen at twice its valuation of the time in the format and appearance of the Bloomberg business news service had been posted [27]. This raises the issues of protecting the authenticity and identification of the origin of data, as well as means of protection for the integrity and authenticity of collections of data from multiple sources that may be assembled not only dynamically but in response to personalized configurations of individual users.

References

[1] Primo Braga, C. A., C. Fink, and C. Paz Sepulveda, *Intellectual Property Rights and Economic Development*, technical report, The World Bank, Washington D.C., 2000.

[2] McCray, W. P., *Glassmaking in Renaissance Venice: The Fragile Craft*, Aldershot, U.K.: Ashgate Publishing Company, 1999.

[3] Goldstein, P., *International Legal Materials on Intellectual Property*, New York: Foundation Press, 2000.

[4] Correa, C. M., *Intellectual Property Rights, the WTO and Developing Countries*, London: Zed Books, 2000.

[5] Berne Convention for the Protection of Literary and Artistic Works of September 9, 1886, completed at Paris on May 4, 1896, revised at Berlin on November 13, 1908, completed at Berne on March 20, 1914, revised at Rome on June 2, 1928, at Brussels on June 26, 1948, at Stockholm on July 14, 1967, and at Paris on July 24, 1971, and amended on September 28, 1979. WIPO Database of Intellectual Property Legislative Texts, September 1979. Paris Act of July 24, 1971, as amended on September 28, 1979.

[6] Lawlor, H. J., "The Cathach of St. Columba," *Proceedings of the Royal Irish Academy*, Vol. 33, No. 11, 1916, pp. 241–443.

[7] Patterson, L. R., and S. W. Lindberg, *The Nature of Copyright: A Law of Users' Rights*, Athens, GA: University of Georgia Press, 1991.

[8] Kant, I., "Von der Unrechtmäßigkeit des Büchernachdrucks," *Berlinische Monatsschrift*, 1:403–417, 1785.

[9]   World Intellectual Property Organization Copyright Treaty, WIPO Database of Intellectual Property Legislative Texts, December 1996. (Adopted by Diplomatic Conference on December 20, 1996.)

[10] Thierer, A., and C. W. Crews, Jr., (eds.), *Copy Fights: The Future of Intellectual Property in the Information Age*, Washington D.C.: The Cato Institute, 2002.

[11] International Federation of the Phonographic Industry, *IFPI Music Piracy Report*, June 2001.

[12] Gomes, L., "Web Piracy Is Hitting Hollywood Sooner Than the Studios Thought," *The Wall Street Journal*, February 17, 2000.

[13] Gates, W. H., III, "An Open Letter to Hobbyists," *Computer Notes*, December 1976.

[14] Freiberger, P., and M. Swaine, *Fire in the Valley: The Making of the Personal Computer*, 2nd ed., New York: McGraw-Hill, 2000.

[15] ISO/IEC Joint Technical Committee 1 Subcommittee 29 Working Group 11, *Information Technology: Coding of Moving Pictures and Associated Audio for Digital Storage Media at up to About 1.5 Mbit/s Part 3: Audio*, ISO/IEC 11172-3, 1993.

[16] ISO/IEC Joint Technical Committee 1 Subcommittee 29 Working Group 11, *Coding of Moving Pictures and Audio. Overview of the MPEG-4 Standard*, ISO/IEC N3156, 1993.

[17] Anderson, R., and M. Kuhn, "Tamper Resistance—A Cautionary Note. *Proceedings of the 2nd USENIX Workshop on Electronic Commerce*, Oakland, CA: USENIX, November 1996, pp. 1–11.

[18] Ansberry, C., "Alterations of Photos Raise Host of Legal, Ethical Issues," *The Wall Street Journal*, Vol. 26, January 26, 1989, p. 131.

[19] Power, M., "Can You Believe Your Eyes?" *The Washington Post*, April 26, 1989.

[20] NPPA, *The Ethics of Photojournalism*, technical report, Durham, NC: The National Press Photographers Association, 1990.

[21] Kurtz, H., "Newsday's Phony Skater's Waltz," *The Washington Post*, February 18, 1994.

[22] Ritchin, F., "Photography's New Bag of Tricks," *The New York Times*, November 4, 1984, p. 49.

[23] U.S. Senate Committee on the Judiciary, *Communist Forgeries: Hearing Before the Subcommittee to Investigate the Administration of the Internal Security Act and Other Internal Security Laws of the Committee of the Judiciary*, Testimony of Richard Helms, Assistant Director, Central Intelligence Agency, Washington, D.C.: GPO, June 1961.

[24] Brugioni, D., "Spotting Photo Fakery," *Studies in Intelligence*, Vol. 13, No. 1, 1969, pp. 57–67.

[25] Jaubert, A., *Making People Disappear*, Washington D.C.: Pergamon-Brassey's, 1986.

[26] Brugioni, D. A., *Photo Fakery: The History and Techniques of Photographic Deception and Manipulation*, London: Brassey's, 1999.

[27] Evgeniy Gabrilovich and Alex Gontmakher, "The Homograph Attack," *Communications of the Association for Computing Machinery*, Vol. 45, No. 2, February 2002, p. 128.

Contents

Digital watermarking

2.1 Rationale

Steganography and watermarking describe methods to embed information transparently into a carrier signal. Steganography is a method that establishes a covered information channel in point-to-point connections, whereas watermarking does not necessarily hide the fact of secret transmission of information from third persons. Besides preservation of the carrier signal quality, watermarking generally has the additional requirement of robustness against manipulations intended to remove the embedded information from the marked carrier object. This makes watermarking appropriate for applications where the knowledge of a hidden message leads to a potential danger of manipulation. However, even knowledge of an existing hidden message should not be sufficient for the removal of the message without knowledge of additional parameters such as secret keys. A crucial feature of digital watermarking is to hide the additional information not in distinguished locations in a specific media format such as the header of a file—which could be lost during transformation into another presentation format—but directly in the signal to be marked itself. This requires a certain perceptual threshold allowing the insertion of additional information and hence distortions of the carrier signal without incurring unacceptable perceptual degradation of the original carrier signal. This implies that marking of executable program code will be difficult, since any arbitrary modification to the bit stream could destroy the functioning of the program, while changes not affecting the semantics of the program can be removed easily through a normalization process. Watermarking systems are therefore context-specific, that is, the algorithms must be designed with respect to the media type of the data to be watermarked. In this sense, watermarking represents a specific application of steganographic techniques. Specifically, the additional requirement for robustness of digital watermarks against attacks or manipulations during the data processing entails a lower data rate of the watermarking methods compared to steganographic algorithms.

2.2   Digital watermarking and cryptography

Most content protection mechanisms (for exceptions, see Chapter 8) rely on cryptological (cryptographical or steganographical) means for the provision of functionality. These mechanisms serve one or more of the requirements in definitions for confidentiality through anonymity that are commonly sought for in information security.

In the following, it is assumed that readers are familiar with basic applied cryptography (see [1–14]).

2.2.1   Steganography

The distinction between cryptography and steganography was not made ab initio; the term *steganographia* first appears in a manuscript by Johannes Trithemius that was started in 1499 but never completed and did not yet make the distinction between the two terms [15]; this was still the case in a book by Caspar Schott published in 1665, which largely contained cipher systems [16]. The narrower definition of cryptography is due to the founder of the Royal Society, John Wilkins, who defined the term as *secrecy in writing* [2, 17].

Steganography is the study of techniques for hiding the existence of a secondary message in the presence of a primary message. The primary message is referred to as the *carrier signal* or *carrier message*; the secondary message is referred to as the *payload signal* or *payload message*. Classical steganography (i.e., steganographic techniques invented prior to the use of digital media for communication) can be divided into two areas, *technical steganography* and *linguistic steganography*. The classification of the various steganographic techniques is shown in Figure 2.1 and described briefly in the following section.

Steganography itself offers mechanisms for providing confidentiality and deniability; it should be noted that both requirements can also be satisfied solely through cryptographic means.

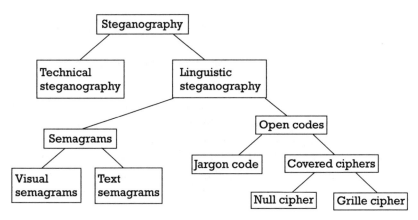

Figure 2.1 Classification of steganographic techniques. (Adapted from [17].)

2.2.1.1 Technical steganography

Technical steganography involves the use of technical means to conceal the existence of a message using physical or chemical means. Examples of this type of steganography include invisible inks, which have been known since antiquity [2, 18], or, more recently, photomechanical reduction resulting in so-called microdots that permit the reduction of a letter-sized page onto an area of photographic film no larger than the dot at the end of this sentence. Although credit is commonly assigned to German intelligence (who made extensive use of microdots prior to and during World War II), the first documented microdots, albeit some 5,000 mm^2 in size, were used by the French during the Franco-Prussian war of 1870–1871 and primarily intended for transportation by pigeons [19, 20]. While a fascinating subject in itself, the discussion here is not concerned with it.

2.2.1.2 Linguistic steganography

Linguistic steganography itself can be grouped into two categories, *open codes* and *semagrams*. The latter category also encompasses *visual semagrams*. These are physical objects, depictions of objects or other diagrams with an ostensibly innocuous purpose in which a message is encoded. Examples of semagrams include the positioning of figures on a chessboard or the drawing of dancing men in Doyle's "The Adventure of the Dancing Men" [21], shown in Figure 2.2.

Text semagrams are created by modifying the appearance of a text in such a way that the payload message is encoded. A number of techniques have been

Figure 2.2   Sherlock Holmes' dancing men semagram.

devised for this purpose for both manuscripts and printed books. Examples include the interruption of connecting lines in longhand manuscripts, the use of slightly different fonts from the main body of the text to encode the payload message [2, 22], or the punctuation of characters either representing directly or by virtue of another encoding such as the distance between characters using, for example, pinpricks or markings in (invisible) ink over the selected characters [20].

The category of open codes is characterized by embedding the payload signal in such a way in the carrier signal that the carrier signal itself can be seen as a legitimate message from which an observer may not immediately deduce the existence of the payload signal. The most obvious use of open codes occurs in the use of codewords such as the *Ave Maria* code by Trithemius [23], where individual characters, words, or phrases are mapped onto entities of the carrier signal. Occasionally the use of such codes is unintentional, as is made evident by the term commonly used for such open codes, *jargon code*.

Conversely, *cue codes* use a possibly elaborate carrier message to signal the occurrence of an event whose semantics have been prearranged. One of the most frequently cited examples is the cue code with which World War II Japanese diplomats were to be notified of impending conflict. In this code, "HIGASHI NO KAZE AME" ("east wind, rain") signified pending conflict with the United States, while "KITANO KAZE JUMORI" ("north wind, cloudy") indicated no conflict with the U.S.S.R. and "NISHI NO KAZE HARE" ("west wind, clear") with the British Empire[1] [24]. Unlike jargon codes, which lead to atypical language that can be detected by an observer, cue codes are harder to detect provided that their establishment has not been compromised.

Another mechanism, commonly referred to as *grille ciphers,* is based on the imposition of a grid known only to the communicating parties onto a message consisting of characters or words commonly attributed to Girolamo Cardano and reading the elements left uncovered by the grille in a predefined order [17, 25]. It was still in active use by the German army in 1914 [26].

---

1. The information containing this code was dispatched on November 26 from the Japanese Foreign Ministry to diplomatic and consular officials in enciphered form; while the message could be deciphered by the Navy Department on December 5, 1941, it did not result in adequate countermeasures.

A variation on the theme of the grille is the use of *null ciphers*. The payload message is embedded as plain text (hence the null cipher) within the carrier message. The communicating parties must prearrange a set of rules that specify the extraction of the payload message (occasionally also found in literature, an acrostic construct arranges verses in such a way that initial or final letters spell out another word; more elaborate versions were used for steganographic purposes). The payload message may also be subject to encoding prior to embedding in the carrier message; this technique was used by Johann Sebastian Bach in a number of works; the canonical example here is *Vor deinem Thron* (BWV 541), which contains a sequence where g occurs twice, a once, h three times, and c eight times; while this and other isopsephic encodings have been found [27], this has also been the subject of debate [28].

2.2.2   Digital watermarking

The original purpose of steganographic mechanisms has been information hiding. The techniques and extensions to them based on the possibilities provided by the digital representation of media, however, suggest another application area, namely the protection of a marking against removal. Analogous to a mechanism for the analog, paper domain [29, 30], this was termed *digital watermarking*.

This implies additional properties not present for steganography, since in digital watermarking one must assume that an adversary has knowledge of the fact that communication is taking place. These requirements are that the embedded signal must be redundant so as to provide robustness against selective degradation and removal and it must be embedded in such a way that it cannot be replaced by a fraudulent message or removed entirely; the latter goal is typically achieved by requiring knowledge of a secret for embedding and removal.

2.3   First generation approaches

During the early to mid-1990s, digital watermarking attracted the attention of a significant number of researchers after several early works that may also be classified as such [31]. Since then the number of publications has increased exponentially to several hundred per year [30]. It started from simple approaches presenting the basic principles to sophisticated algorithms using results from communication theory and applying them to the watermarking problem.

2.3.1 Basic principles of watermarking

Since this research field is still relatively young and has contributors from several disciplines with varying traditions, the terminology used is still quite diverse. This section provides a formal introduction to watermarking systems and the terms used in this context for their presentation.

Formal description of watermarking   The basic principle of current watermarking systems are comparable to symmetric encryption as to the use of the same key for encoding and decoding of the watermark. Each watermarking system consists of two subsystems: a watermarking encoder and a respective decoder. Formally, a watermarking system can be described by a tuple $\langle \mathcal{O}, \mathcal{W}, \mathcal{K}, E_K, D_K, C_\tau \rangle$, where $\mathcal{O}$ is the set of all original data, $\mathcal{W}$ the set of all watermarks, and $\mathcal{K}$ the set of all keys. The two functions

$$E_K : \mathcal{O} \times \mathcal{W} \times \mathcal{K} \longrightarrow \mathcal{O} \tag{2.1}$$

$$D_K : \mathcal{O} \times \mathcal{K} \longrightarrow \mathcal{W} \tag{2.2}$$

describe the embedding and detection process, respectively. The comparator function

$$C_\tau : \mathcal{W}^2 \longrightarrow \{0, 1\} \tag{2.3}$$

compares the extracted with the really embedded watermark using the threshold τ for comparison. The input parameters of the embedding process are the carrier object (or original $\mathbf{c_o}$), the watermark $\mathbf{w}$ to be embedded, as well as a secret or public key K:

$$E_K(\mathbf{c_o}, \mathbf{w}) = \mathbf{c_w} \tag{2.4}$$

The output of the encoder forms the marked data set (see Figure 2.3).

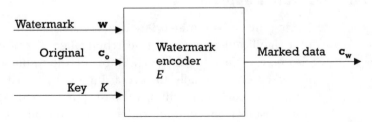

Figure 2.3 Generic watermark encoder.

In the detection process, the marked and possibly manipulated data set $\hat{\mathbf{c}}_{\mathbf{w}}$, the original $\mathbf{c_o}$, the watermark $\mathbf{c_w}$, and the key K used during the embedding process form the maximal set of input parameters (see Figure 2.4).

The various types of watermarking systems differ in the number of input parameters in the reading process (see Section 2.3.2). The extracted watermark $\hat{\mathbf{w}}$ differs in general from the embedded watermark $\mathbf{w}$ due to possible manipulations. In order to judge the correspondence of both watermarks, the comparator function C_τ compares the suspected watermark with the retrieved one against a threshold τ:

$$C_\tau(\hat{\mathbf{w}}, \mathbf{w}) = \begin{cases} 1, & c \geq \tau \\ 0, & c < \tau \end{cases} \tag{2.5}$$

The threshold τ depends on the chosen algorithm and should in a perfect system be able to clearly identify the watermarks. This formal analysis of the watermarking systems can also be used to develop a geometric interpretation of the watermarking algorithms as shown in [30].

2.3.2   Terminology

2.3.2.1   Types of watermarks

▸ *Robust watermarks* are designed to resist against heterogeneous manipulations; all applications presupposing security of the watermarking systems require this type of watermark.

▸ *Fragile watermarks* are embedded with very low robustness. Therefore, this type of watermark can be destroyed even by the slightest manipulations. In this sense they are comparable to the hidden messages in steganographic methods. They can be used to check the integrity of objects.

▸ *Public* and *private watermarks* are differentiated in accordance with the secrecy requirements for the key used to embed and retrieve markings.

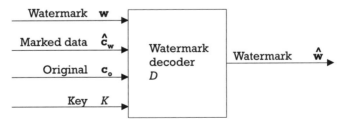

Figure 2.4   Generic watermark decoder.

According to the basic principle of watermarking, the same key is used in the encoding and decoding process. If the key is known, this type of watermark is referred to as public, and if the key is hidden, as private watermarks. Public watermarks can be used in applications that do not have security-relevant requirements (e.g., for the embedding of meta information).

- *Visible* or *localized watermarks* can be logos or overlay images in the field of image or video watermarking. Due to the implicit localization of the information, these watermarks are not robust.

Besides the various types of watermarks, four different watermarking systems are classified according to the input and output during the detection process. Using more information at the detector site increases the reliability of the whole watermarking system but limits the practicability of the watermarking approach on the embedder side.

The side information in the detection process can be the original c_o and the watermark w itself (see Figure 2.4). Therefore, four permutations of side information requirements are possible.

2.3.2.2 Watermarking systems

- *Nonblind watermarking* [2] systems require at least the original data in the reading process. We can further subdivide this type of system depending on whether or not the watermark is needed within the decoding process.

  Type I systems detect the watermark of the potentially manipulated data set by means of the original:

$$D_K(\hat{c}_w, c_o) = \hat{w} \qquad (2.6)$$

  Type II systems additionally use the watermark and therefore represent the most general case:

$$D_K(\hat{c}_w, c_o, w) = \hat{w} \quad \text{and} \quad C_\tau(\hat{w}, w) = \begin{cases} 1, & c \geq \tau \\ 0, & c < \tau \end{cases} \qquad (2.7)$$

  These systems answer the question: Is the watermark w embedded in the dataset $\hat{c}_w$? In this way the information content of the

---

2. The term *private watermarking* is also used, which can lead to confusion, given the previously introduced terms, *public* and *private watermarks*.

watermark is 1 bit. By using further information, the robustness of these watermarking methods is in general increased.

▶ Different from the above method, *semiblind watermarking* does not use the original for detection:

$$D_K(\hat{\mathbf{c}}_{\mathbf{w}}, \mathbf{w}) = \hat{\mathbf{w}} \quad \text{and} \quad C_\tau(\hat{\mathbf{w}}, \mathbf{w}) = \begin{cases} 1, & c \geq \tau \\ 0, & c < \tau \end{cases} \qquad (2.8)$$

This is essential in applications where access to the original is not practical or possible. Semiblind watermarking methods can be used for copy control and copyright protection.

▶ *Blind watermarking*[3] is the biggest challenge to the development of a watermarking system. Neither the original nor the watermark are used in the decoding process:

$$D_K(\hat{\mathbf{c}}_{\mathbf{w}}) = \hat{\mathbf{w}} \qquad (2.9)$$

This is necessary in applications in which *n* bits of information must be read out of the marked data set $\hat{\mathbf{c}}_{\mathbf{w}}$ as, for example, during the pursuit of illegally distributed copies.

2.3.3   First methods

Watermarking can be considered as communication of the watermark over a channel consisting of the original work to be watermarked. Therefore, a natural approach in development of conceptual models for watermarking is to study the similarities between communication models and corresponding watermarking algorithms. Both models transmit data from an information source (the watermark) to a destination (the user or another system).

The typical model of communication consists of several blocks (shown in Figure 2.5). This model was introduced by Shannon in his landmark 1948 paper [32]. The source message **m** is transformed via a source encoder into a sequence of binary digits **u** representing the encoded source as an information sequence. The process is performed in order to minimize the number of bits representing the source output and to enable the nonambiguous reconstruction of the source from the information sequence [33].

The *channel encoder* transforms the information sequence **u** into an *encoded sequence* **v** called a *code word*.

---

3. Also called *public* or *oblivious* watermarking which, again, can lead to confusion with the previous term, *public watermarks*.

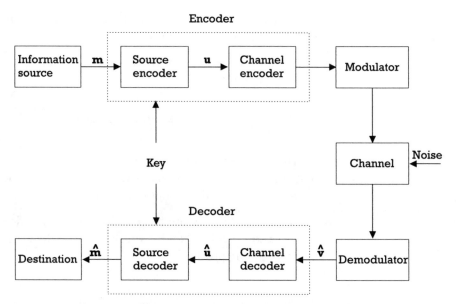

Figure 2.5 Basic communication model for secure transmission.

In order to transmit the discrete symbols over a physical channel, a *modulator* transforms each symbol of the encoded sequence **v** into a form suitable for transmission [34].

During transmission over the channel, the transformed sequence is distorted by noise. The different forms of noise that can disturb the transmission are driven by the channel characteristics. On the receiver site, the *demodulator* processes the transmitted sequence and produces an output **v̂** consisting of the counterpart of the encoded sequence. Corresponding to the encoder, the *channel decoder* transforms the output of the demodulator into a binary sequence **û**, which is an estimation of the true sequence being transmitted. In a perfect channel, the estimated sequence **û** would be a copy of the true sequence *u*. Carefully designed source encoders can reduce coding errors originated from the disturbance by the noise of the channel [33]. The last step is performed by the *source decoder*, which transforms the decoded sequence **û** into an estimate of the source output sent to the destination.

The different types of communication channels can be categorized by the type of noise introduced during the transmission and how the noise is applied to the signal [34].

Besides the channel characteristics, the transmission can be further classified according to the security it provides against active attacks trying to disable communication and against passive attacks trying to monitor the communication (read the transferred messages).

The defense against the attacks is based on:

‣ Spread-spectrum techniques trying to prevent active attacks;

‣ Cryptography encrypting the messages in order to ensure privacy.

Digital watermarking and spread-spectrum techniques try to fulfill similar security requirements in preventing active attacks like jamming the communication between different communicating parties. Spread-spectrum technologies establish secrecy of communication by performing modulation according to a secret key in the channel encoder and decoder (see Figure 2.5).

A watermarking model based on communication consists of the same basic blocks as the communication model with different interpretations. There is a direct correspondence between the watermark embedder/detector and the channel encoder/decoder—including the modulation/demodulation blocks—respectively. The message to be transmitted is the watermark itself. The additional requirement of secure transmission of the signal over the channel requires the usage of a secret key in the encoding and decoding procedure (see Section 2.3.1).

The channel characteristics can be modeled by:

‣ The cover object representing the channel carrying the watermark;

‣ The kind of noise introduced by the different processing that can happen during transmission of the watermarked object. This additional processing may be anticipated manipulations or intentional attacks.

The encoding block of the watermark embedder encodes the watermark message $\mathbf{m}$ into a coded sequence $\mathbf{v}$. During the modulation, the sequence $\mathbf{v}$ is transformed into a physical signal, the watermark signal $\mathbf{w}$, that can be transmitted over the channel. The difference between the marked and original cover object—which actually forms the added watermark—will essentially have the same digital representation as the original data set. For example, in the case of an audio file, the added watermark will be a signal with the same sample rate and bit resolution as the cover track. At the watermark detector site, the possible distorted marked object is demodulated into $\hat{\mathbf{v}}$, which is a distorted version of the coded sequence $\mathbf{v}$. The watermark message $\hat{\mathbf{m}}$ is obtained by means of the watermark message decoder (see Figure 2.6) from $\hat{\mathbf{v}}$.

Analogous with the basic communication system, the encoder has to perform three steps:

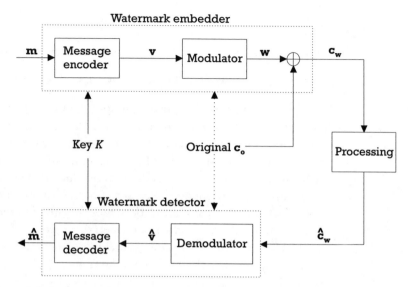

Figure 2.6 Basic watermark communication model.

1. Encode the message into a coded sequence using a secret key.

2. Modulate the coded sequence into a physical representation according to the channel respectively the cover object.

3. Add the modulated sequence to the cover object to produce the watermarked object.

In order to classify existing digital watermarking systems, the basic building blocks of watermark embedders/detectors must be examined in closer detail.

The first approaches implement watermark encoders by addition of the generated watermark pattern without considering the channel characteristics respectively the cover object c_0. Methods from this first generation predefine a set of significant components in a so-called *embedding domain* for watermark embedding based on some heuristic criteria [35, 36]. As depicted in Figure 2.7, this usually involves a transformation into another signal representation[4] where the alteration of the preselected carrier components is performed. A number of methods are working in the Fourier domain using the low to middle frequency range as components to embed the watermark [37].

---

4. Typical embedding domains used for modification are the Fourier, wavelet, and Cepstrum domains.

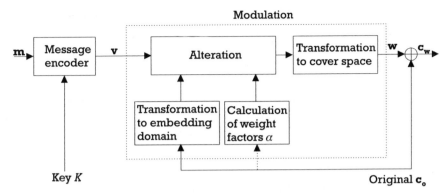

Figure 2.7 Basic watermark embedder.

From experiments it is known that these are perceptually significant components [38, 39], resulting in a compromise between the quality of the watermarked object and the robustness of the embedded watermark. To enable the adjustment between the two opposite requirements of perceptual visibility and robustness of the watermark, a vector α of weight factors is calculated. The very first algorithms [35] have used a vector with equal elements $\alpha = \{\alpha[i]\}_{i=1}^{N}$, $\alpha[i] = \alpha_{\text{const}}$, $\forall i$ to control the embedding strength respectively the power of the embedded watermark (see Figure 2.7). The obvious disadvantage of such simple methods is the missing relation between the cover object and the embedded watermark, resulting in a possible loss of fidelity if an overall weight factor is used without considering local variations of the cover object respectively the channel. Furthermore, optimization of the robustness of the embedded watermark with regard to the actual cover object is not taken into account. From this point of view, little or no information about the specific cover object is taken into account to improve both the embedding and detection procedure. More advanced algorithms [40–44] optimize the quality by investigating the cover object to calculate perceptual thresholds and the corresponding weight vector α. These so-called perceptual thresholds are derived from perceptual models [45–47] for different media types. This information is used in the modulation block of the encoder to shape the added pattern to ensure maximum quality. Therefore, the added pattern is a function of the cover object. These methods use the perceptual thresholds to optimize for quality, usually neglecting the effect on the robustness of the watermark.

The decoders perform two steps to retrieve the watermark:

1. The obtained signal is demodulated to obtain a message pattern.

2. The message pattern is decoded with the decoding key in order to retrieve the embedded watermark.

The demodulation can be performed in different ways. If the original object is available to the decoder, it can be subtracted from the received signal, which is the watermarked object, to obtain the received noisy watermark pattern (see Figure 2.8).

Demodulation in this way is done in nonblind watermarking systems (see Section 2.3.2). Other approaches use data reduction functions to cancel out the effect of the addition of the cover object. This can be done by approximating the original in the detection procedure before subtracting it from the watermarked object.

In the decoding procedure, the coded sequence $\hat{v}$ of the watermark has to be extracted by the sequence extractor. In early approaches, the same predefined carrier components in the embedding space are used for sequence retrieval as those in the embedding step. The watermark message $\hat{m}$ is decoded in the watermark message decoder from $\hat{v}$ by means of the secret key.

2.3.3.1 LSB coding

One of the first techniques investigated in the watermarking field, as in virtually all media types, is the so-called *least significant bit* (LSB) method of encoding. It is based on the substitution of the LSB of the carrier signal with the bit pattern from the watermark noise. The bits are embedded in certain representation values, such as pixels. The decoder in turn is able to retrieve the watermark if it has the knowledge of the representation values used for embedding the individual bits.

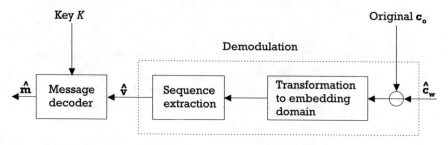

Figure 2.8 Basic watermark detector.

The watermark encoder uses a subset $c_{oj}[1], \ldots, c_{oj}[l(c_{oj})]$[5] of all available carrier elements c_o chosen by the secret key. The substitution operation $c_{oj}[i] \rightleftharpoons m[i]$ on the LSBs is performed on this subset.[6] The reading process retrieves the values of this bits. Therefore, the decoder needs all the carrier elements that were used during the embedding process. Generally, $l(c_o) \gg l(m)$; therefore, the robustness can be improved by a repeated embedding of the watermark. The random selection of the elements for embedding and the changing of the LSBs generate noise with low power and a constant power density. The perception of this noise depends on the perceptual threshold of the original carrier object and therefore depends on its content.

The main advantage of this method is its high payload, whereas the main disadvantage lies in its low robustness, due to the fact that random changes of the LSBs destroy the coded watermark. It is, for example, extremely unlikely that a LSB encoded watermark will survive a digital-to-analog and subsequent analog-to-digital conversion. The characteristics of the LSB methods limit their applicability to steganographic scenarios and requires a purely digital environment.

2.3.3.2   Spread-spectrum techniques

Spread-spectrum communication encompasses a number of signaling techniques in which the transmitted bandwidth is significantly larger than required by the data rate and the transmitted bandwidth is determined by a function independent of the message that is known to both sender and receiver. Originally intended to increase resistance to jamming and lowering the probability of interception, its properties are also desirable in a number of other applications [48, 49] and have been applied to the problem of digital watermarking early on [35].

The invention of spread-spectrum communication is commonly attributed to Kiesler[7] and Antheil, who devised a mechanism for sending guidance information to torpedoes believed to be resistant to jamming by means of sending the information distributed across a wide frequency band in a predetermined pattern known to both receiver and sender,[8] that is, "frequency hopping" [50]. However, even more advanced concepts had emerged earlier and in other locations such as in the RADAR system patented by Guanella

---

5. An additional index in c_{oj} denotes a subset of the signal and $l(c_{oj})$ denotes the length of the vector c_{oj}.
6. The substitution can be performed on more than 1 bit of the carrier element.
7. Better known under the name Hedy Lamarr.
8. Antheil, a pianist, proposed the use of punch tape found in player pianos for this purpose.

in 1938 in which signal correlation and autocorrection were already present and the 1940 patent by Kotowski and Dannehl [51] describing an encryption device based on the combination of broadband noise with the payload signal [52].

Model   Within the terminology commonly used in the spread-spectrum context, the carrier signal can be considered as a jammer interfering with the payload signal carrying the watermark information (see also Figure 2.6).

Assuming that the carrier signal is modeled by N waveforms, and a payload signal that is modeled by $D \leq N$ orthonormal basis functions and given $\{\varphi_k(t); 1 \leq k \leq N\}$ orthonormal bases spanning the signal space such that the carrier signal $c(t)$ is modeled by (2.10) in an N-dimensional space and that the signal is without loss of generality variable over time t:

$$c(t) = \sum_{k=1}^{N} c_k \varphi_k(t); \quad 0 \leq t \leq T \tag{2.10}$$

such that the total energy in the carrier signal is given by (2.11) (the validity of this assumption must be verified for each type of signal) [53].

$$\int_0^T c^2(t)dt = \sum_{k=1}^{N} c_k^2 \triangleq E_c \tag{2.11}$$

The payload signal can then without loss of generality be modeled by D equiprobable and equienergy orthogonal signals according to (2.12):

$$s_i(t) = \sum_{k=1}^{N} s_{ik} \varphi_k(t); \quad 1 \leq i \leq D, 0 \leq t \leq T \quad \text{where}$$

$$s_{ik} = \int_0^T s_i(t)\varphi_k(t)dt \quad \langle \phi_l, \phi_m \rangle = \int_0^T \varphi_l(t)\varphi_m(t)dt = \delta_{lm} \triangleq \begin{cases} 1 & l = m \\ 0 & l \neq m \end{cases}$$

$$\tag{2.12}$$

The average energy of each payload signal can then be given in (2.13), with $\bar{x}$ representing the expected value over the statistic ensemble.

$$\int_0^T \overline{s_i^2}(t)dt = \sum_{k=1}^{N} \overline{s_{ik}^2} \triangleq E_s; \quad 1 \leq i \leq D \tag{2.13}$$

The combined carrier and payload signals $c(t) + s(t)$ are correlated by the receiver with the known signal. Assuming as above that the energy of the carrier signal is bounded, the average processing gain is equal to the ratio of dimensionality of the signal space to the subspace taken by the payload signal, independent of the signal energy distribution of the carrier signal [54].

This section can only provide a brief introduction to spread-spectrum mechanisms; many of the more advanced concepts developed in the context of RADAR A/J, particularly those related to jamming resistance, are applicable to digital watermarking [54–59].

Recovery   The recovery of the spread-spectrum carrier can be accomplished using three general techniques [52]:

1. In *transmitted reference* (TR) schemes, detection is accomplished by sending the carrier signal in two versions, one modulated and one unmodulated and processing both signals through a correlation detector.

2. *Stored reference* (SR) schemes require sender and receiver to possess the means to generate the same pseudorandom signal. In addition, the receiver must have the ability to adjust the carrier generator for synchronization of the output with the incoming carrier; the actual detection is the same as in TR schemes.

3. *Matched filtering* (MF) providing a wideband pseudorandom impulse response can be used at the receiver side to recover the spread-spectrum signal.

Of interest for digital watermarking are mainly TR schemes, which can be considered a form of nonblind watermarking; the more common watermarking mechanisms employ SR schemes, which are applicable to both semiblind and blind watermarking.

2.3.3.3 Patchwork technique

The data to be watermarked is separated into two distinct subsets. One feature of the data is chosen and modified in opposite directions in both subsets. For example, sample values are labeled as belonging to subset $\mathcal{A}$ or $\mathcal{B}$. In subset $\mathcal{A}$ the data values are increased, while in subset $\mathcal{B}$ their values are decreased by a certain amount Δ. The separation of the samples is the secret used in the embedding and detection step. The watermark can be easily detected if the data satisfies some statistical properties. This technique was first presented by Bender et al. [36]:

Let $N = N_A = N_B$ equal the size(s) of the subsets and Δ the amount of the change of individual samples. Let $\mathbf{a}[i]$ equal the sample data at position i of subset $\mathcal{A}$ whose values are increased. Let $\mathbf{b}[i]$ equal the sample data in subset $\mathcal{B}$. The difference of the sample values can be written as

$$S = \frac{1}{N_A} \sum_{N_A} \mathbf{a}[i] - \frac{1}{N_B} \sum_{N_B} \mathbf{b}[i] = \frac{1}{N} \sum_N (\mathbf{a}[i] - \mathbf{b}[i])$$

The expectation of the difference can be calculated as

$$E\{S\} = \begin{cases} 2\Delta & \text{for watermarked data} \\ 0 & \text{for unwatermarked data}^9 \end{cases}$$

A test statistic can be defined that is compared against a threshold value. Cox et al. [30] showed that although this patchwork technique appears totally different from the correlation-based approach of spread-spectrum techniques, it is similar to linear correlation, which can be shown easily: Instead of calculating the differences of the two sums, one can correlate data values against a specific pattern. This pattern consists either of 1 or -1 depending on the group the corresponding data values belong to.

2.3.3.4 Quantization index modulation

Watermarking by *quantization index modulation* (QIM) proposed by Chen and Wornell [60] is based on a set of N-dimensional quantizers. The message $\mathbf{m}$ that should be transmitted is the index for the quantizer used for quantizing the host-signal vector $\mathbf{c_o}$. While retrieving the hidden information, one evaluates a distance metric to all quantizers. The index of the quantizer with the smallest distance contributes to the message $\mathbf{m}$. To reduce distortion, the distortion constraint has to be fulfilled: $E_K(\mathbf{c_o}, \mathbf{m}) = \mathbf{c_m} \approx \mathbf{c_o}, \forall \mathbf{m}$. To increase robustness, the reconstruction values of the different quantizers must have a maximum distance. An example of the two-dimensional case ($m = 2$) is shown in Figure 2.9.

In an extensive analysis, Chen and Wornell showed that QIM methods are near-optimal for Gaussian channels.

2.4 Beyond the first generation

In contrast to the initial approaches, more advanced algorithms take additional information into account in the encoding process. The side information used in the watermark embedder concerns the channel characteristics respectively the cover object $\mathbf{c_o}$. The cover information can be used at several

---

9. The subsets N_A and N_B are chosen from the same population N. Therefore, $E\{A\} \approx E\{B\} \approx E\{A \cup B\}$ and $E\{A\} - E\{B\} \approx 0$ for unwatermarked data.

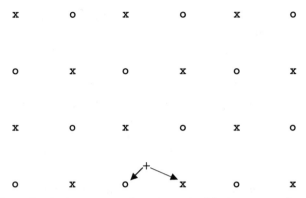

Figure 2.9 Example of using two quantizers for embedding a watermark: x and o belong to different quantizers. + indicates the current value that should be quantified. The result depends on the chosen quantizer.

stages of the embedding process. The preselection of the carrier elements is not specified in advance according to some heuristic criteria driven by basic perceptual facts using a fixed frequency range, but rather by different algorithmic requirements. A feature extraction block is often combined with the transformation in the embedding domain to select carrier components that are more appropriate for watermark embedding than others.

Selection criteria are, for example, reduced correlation between the cover object and the watermark signal to be embedded, or higher possible alterations of components according to the perceptual thresholds to improve the robustness. This information can be used in the alteration block in Figure 2.10 to optimize between quality on the one hand and robustness on the other hand according to application requirements.

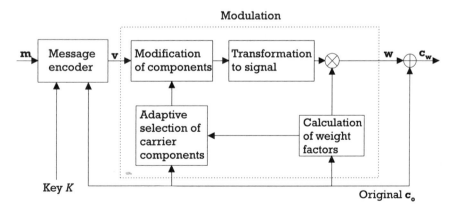

Figure 2.10 Advanced watermark embedder.

Moreover, the information of the cover object can be used in order to adjust the coded sequence with respect to the message carrier respectively the cover object. A search through a set of code vectors can be performed in order to embed the code vector closest to the cover object. The set of all code vectors is tested at the decoder site in order to identify the message corresponding to the code vector found. Research in using side information (the cover object) in the watermark encoder includes the so-called *dirty paper* channel studied by Costa [61]. By using perceptual models and feature extraction procedures at the watermark detector site, one can also improve the detection reliability of the watermark.

Feature extraction has to be performed in order to select the corresponding regions or components used during watermark embedding. Perceptual models can be applied at the detector site to calculate the perceptual weight factors. Depending on the input of the model, which can be the original or the marked data set, the weight factors are exactly the same as in the embedding process or approximations thereof. Since one of the main requirements of watermarking is perceptual transparency of the marked data set, one can assume that the calculation of the masking threshold leads to results similar to those in the original case. Therefore, the approximations of the weight factors derived from the marked cover object should lead to satisfactory results. To improve detection performance, knowledge of the perceptual characteristics can be used by applying inverse weighting to the carrier components in order to remove the perceptual shaping which can be regarded as a noisy process during embedding (see Figure 2.11).

General methods are used for modulation of the information, taking the cover object or the transmission channel, respectively, into account. Yet to achieve a balance between quality of the watermarked object and robustness of the embedded watermark requires a deep understanding of the characteristics of the carrier data type. Therefore, the algorithms for digital watermarking have to be tailored to the specific media type. Chapters 4 to 6 present watermarking algorithms and techniques for various media types.

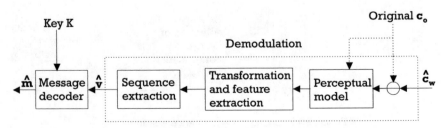

Figure 2.11 Advanced watermark detector.

References

[1] Menezes, A. J., P. C. van Oorschot, and S. A. Vanstone, *Handbook of Applied Cryptography*, Boca Raton: CRC Press, 1996.

[2] Kahn, D., *The Codebreakers: The Comprehensive History of Secret Communication from Ancient Times to the Internet*, 2nd ed., New York: Scribner, 1996.

[3] Deavours, C. A., et al., (eds.), *Cryptology Yesterday, Today, and Tomorrow*, Norwood, MA: Artech House, 1987.

[4] Deavours, C. A., et al., (eds.), *Cryptology: Machines, History & Methods*, Norwood, MA: Artech House, 1989.

[5] Deavours, C. A., et al., (eds.), *Selections from Cryptologia: History, People, and Technology*, Norwood, MA: Artech House, 1998.

[6] Shannon, C. E., "Communication Theory of Secrecy Systems," *Bell System Technical Journal*, Vol. 28, No. 4, October 1949, pp. 656–715.

[7] Kerkhoffs, A., "La Cryptographie Militaire," *Journal des Sciences Militaires*, 9th series, January/February 1883, pp. 5–38, 161–191.

[8] Luby, M., *Pseudorandomness and Cryptographic Applications*, Princeton Computer Science Notes, Princeton: Princeton University Press, 1996.

[9] Rabin, M. O., "Digitalized Signatures," in R. DeMillo et al., (eds.), *Foundations of Secure Computation*, Boston: Academic Press, 1978, pp. 155–168.

[10] Lamport, L., *Constructing Digital Signatures from a One-Way Function*, technical report, Palo Alto, CA: SRI International, October 1979.

[11] Merkle, R. C., "A Certified Digital Signature," in Gilles Brassard, (ed.), *Advances in Cryptology (CRYPTO '89)*, *Lecture Notes in Computer Science*, Vol. 435, Santa Barbara, CA: Springer-Verlag, August 1989, pp. 218–238.

[12] Goldwasser, S., S. Micali, and C. Rackoff, "A Digital Signature Scheme Secure Against Adaptive Chosen-Message Attacks," *SIAM Journal on Computing*, Vol. 17, 1988, pp. 281–308.

[13] ISO/IEC Joint Technical Committee 1 Subcommittee 27: IT Security Techniques. *Information Technology—Security Techniques—Entity Authentication Mechanisms—Part 1: General Model*, 2nd ed., ISO/IEC 9798-1, 1997.

[14] Diffie, W., and M. E. Hellman, "New Directions in Cryptography," *IEEE Transactions on Information Theory*, Vol. 22, No. 60, November 1976, pp. 644–654.

[15] Trithemius, J., *Polygraphiae libri sex. . . Accessit clavis polygraphiae liber unus, eodem authore*, Ioannes Birckmannus & Theodorus Baumius, 1586. (Copies are located at the National Cryptologic Museum, Ft. George G. Meade, MD, and the Hill Monastic Manuscript Library, Bush Center, Saint John's University, Collegeville, MN.)

[16] Schott, C., *Schola steganographia*, Jobus Hertz, 1665.

[17] Bauer, F. L., *Kryptologie: Methoden und Maximen*, 2nd ed., Heidelberg: Springer-Verlag, 1991.

[18] Norman, B., *Secret Warfare*, 2nd ed., Washington D.C.: Acropolis Books, 1996.

[19] Hayhurst, J. D., *The Pigeon Post into Paris 1870–1871*, privately published by the author, 1970.

[20] Polmar, N., and T. B. Allen, *The Encyclopaedia of Espionage*, 2nd ed., New York: Random House, 1998.

[21] Doyle, A. C., "The Adventure of the Dancing Men," *Strand Magazine*, Vol. 26, No. 156, December 1903.

[22] deVigenere, B., *Traicté des chiffres, et secretes manieres d'escrire*, Paris, France, 1586.

[23] Trithemius, J., *Steganographia*, Christophorus Küchlerus, 1676. (A copy is located at the Hill Monastic Manuscript Library, Bush Center, Saint John's University, Collegeville, MN.)

[24] Safford, L. F., *Statement Regarding Winds Message by Captain L. F. Safford, U.S. Navy, Before the Joint Committee on the Investigation of the Pearl Harbor Attack*, Collection of Papers Related to the "Winds Execute" Message, U.S. Navy, 1945, document SRH-210, January 1946. (Located in Record Group 457 at the National Archives and Records Administration, College Park, MD.)

[25] Johnson, N. F., and S. Jajodia, "Steganography: Seeing the Unseen," *IEEE Computer*, Vol. 31, No. 2, February 1998, pp. 26–34.

[26] Gardner, M., *Codes, Ciphers and Secret Writing*, Mineola, NY: Dover Publications, 1972.

[27] Smend, F., *Johann Sebastian Bach bei seinem Namen gerufen: Eine Noteninschrift und ihre Deutung*, Basel, Switzerland: Bärenreiter, 1950.

[28] Tatlow, R., *Bach and the Riddle of the Number Alphabet*, Cambridge, U.K.: Cambridge University Press, 1991.

[29] Katzenbeisser, S., and F. A. P. Petitcolas, (eds.), *Information Hiding: Techniques for Steganography and Digital Watermarking*, Norwood, MA: Artech House, 2000.

[30] Cox, I. J., M. L. Miller, and J. A. Bloom, *Digital Watermarking*, The Morgan Kaufmann Series in Multimedia Information and Systems, San Francisco: Morgan Kaufmann Publishers, 2002.

[31] Holt, L., B. G. Maufe, and A. Wiener, "Encoded Marking of a Recording Signal," U.K. Patent GB 2196167, March 1988. Granted January 1991, lapsed May 1995.

[32] Shannon, C. E., "A Mathematical Theory of Communication," *Bell System Technical Journal*, Vol. 27, No. 3, July/October 1948, pp. 379–423, 623–656.

[33] Lin, S., and D. J. Costello, Jr., (eds.), *Error Control Coding: Fundamentals and Applications*, Prentice-Hall Series in Computer Applications in Electrical Engineering, Englewood Cliffs, NJ: Prentice Hall, 1983.

[34] Proakis, J. G., and D. M. Manolakis, *Digital Signal Processing: Principles, Algorithms and Applications*, 2nd ed., Basingstoke, U.K.: Macmillan Publishing Company, 1992.

[35] Cox, I. J., et al., *Secure Spread Spectrum Watermarking for Multimedia*, Technical Report 95-10, NEC Research Institute, 1995.

[36] Bender, W., et al., "Techniques for Data Hiding," *IBM Systems Journal*, Vol. 35, Nos. 3 & 4, 1996, pp. 313–336.

[37] Koch, E., and J. Zhao, "Towards Robust and Hidden Image Copyright Labeling," in I. Pitas, (ed.), *Proceedings of 1995 IEEE Workshop on Nonlinear Signal and Image Processing*, Neos Marmaras, Greece: IEEE Press, June 1995, pp. 452–455.

[38] Terhardt, E., "Calculating Virtual Pitch," *Hearing Research*, Vol. 1, 1979, pp. 155–182.

[39] Mannos, J. L., and D. J. Sakrison, "The Effects of a Visual Criterion on the Encoding of Images," *IEEE Transactions on Information Theory*, Vol. IT-20, No. 4, July 1974, pp. 525–536.

[40] Swanson, M. D., et al., "Robust Audio Watermarking Using Perceptual Masking," *Signal Processing*, Vol. 66, No. 3, May 1998, pp. 337–355.

[41] Neubauer, C., J. Herre, and K. Brandenburg, "Continuous Steganographic Data Transmission Using Uncompressed Audio," in D. Aucsmith, (ed.), *Information Hiding: Second International Workshop*, Vol. 1525 of *Lecture Notes in Computer Science*, Portland, OR: Springer-Verlag, 1998, pp. 208–217.

[42] Barni, M., et al., "A DCT-Domain System for Robust Image Watermarking," *Signal Processing*, Vol. 66, No. 3, May 1998, pp. 357–372.

[43] Podilchuk, C. I., and W. Zeng, "Image-Adaptive Watermarking Using Visual Models," *IEEE Journal on Selected Areas in Communications*, Vol. 16, No. 4, May 1998, pp. 525–539.

[44] Arnold, M., and K. Schilz, "Quality Evaluation of Watermarked Audio Tracks," in P. W. Wong and E. J. Delp, (eds.), *Proceedings of Electronic Imaging 2002, Security and Watermarking of Multimedia Contents IV*, Vol. 4675, San Jose, CA: SPIE, January 2002, pp. 91–101.

[45] Watson, A. B., "DCT Quantization Matrices Visually Optimized for Individual Images," in J. P. Allebach and B. E. Rogowitz, (eds.), *Proceedings of Human Vision, Visual Processing, and Digital Display IV*, San Jose, CA: SPIE, February 1993, pp. 202–216.

[46] Shlien, S., "Guide to MPEG-1 Audio Standard," in *IEEE Transactions on Broadcasting*, Vol. 40, December 1994, pp. 206–218.

[47] Pan, D., "A Tutorial on MPEG/Audio Compression," *IEEE Multimedia*, Vol. 2, No. 2, 1995, pp. 60–74.

[48] Kahn, D., "Cryptology and the Origins of Spread Spectrum," *IEEE Spectrum*, Vol. 21, No. 9, September 1984, pp. 70–80.

[49]   Cooper, G. R., and C. D. McGillem, *Modern Communications and Spread Spectrum*, New York: McGraw-Hill, 1986.

[50]   Braun, H.-J., "Advanced Weaponry of the Stars," *American Heritage of Invention & Technology*, Vol. 12, No. 4, 1997, pp. 10–16.

[51]   Kotowski, P., and K. Dannehl, "Distance Determining System," U.S. Patent 2 211 132, May 1936.

[52]   Scholtz, R. A., "The Origins of Spread-Spectrum Communication," *IEEE Transactions on Communications*, Vol. 30, No. 5, May 1982, pp. 822–854.

[53]   Wozencraft, J. M., and I. M. Jacobs, *Principles of Communication Engineering*, New York: John Wiley & Sons, 1965.

[54]   Pickholz, R. A., D. L. Schilling, and L. B. Milstein, "Theory of Spread-Spectrum Communictions, A Tutorial," *IEEE Transactions on Communications*, Vol. 30, No. 5, May 1982, pp. 855–884.

[55]   Dixon, R. C., *Spread Spectrum Systems with Commercial Applications*, New York: John Wiley & Sons, 1994.

[56]   Simon, M. K., et al., *The Spread Spectrum Communications Handbook*, 2nd ed., New York: McGraw-Hill, 1994.

[57]   Cook, C. E., et al., *Spread-Spectrum Communications*, New York: IEEE Press, 1983.

[58]   Dillard, R. A., and G. M. Dillard, *Detectability of Spread-Spectrum Signals*, Norwood, MA: Artech House, 1989.

[59]   Wilson, S. G., *Digital Modulation and Coding*, Englewood Cliffs, NJ: Prentice Hall, 1995.

[60]   Chen, B. and G. W. Wornell, "Quantization Index Modulation: A Class of Provably Good Methods for Digital Watermarking and Information Embedding," *IEEE Transactions on Information Theory*, Vol. 47, No. 4, May 2001.

[61]   Costa, M., "Writing on Dirty Paper," *IEEE Transactions on Information Theory*, Vol. 29, No. 3, May 1983, pp. 439–441.

Applications of digital watermarking

While significant portions of this book are dedicated to digital watermarking, the intersection with the application domain of content protection is quite limited in that the efficacy of digital watermarks depends considerably on precisely circumscribed application scenarios and boundary conditions. However, there is a range of application scenarios beyond that of content protection for which digital watermarks are also very suitable, particularly for situations where there exists no adversarial situation.

This chapter concentrates on application scenarios and does not consider threats to the digital watermarks or the mechanisms used for embedding and reading these. Most of the threats specific to digital watermarking are discussed in Chapter 7, while Chapter 8 covers threats to the embedding and retrieval mechanism itself.

3.1 Usage-specific requirements

Digital watermarks are particularly attractive for signals constituting a continuous stream such as audio or video signals.

In case such signals are transmitted in analog form, recovery must be possible from the analog form, presumably at a minimum after the signal has been attenuated, distorted, and transformed in the process of transmission and reproduction.

Particularly in the case of analog video signals with their high bandwidth requirements, the recovery must then either be possible given only a very limited high-fidelity recording of the original signal, or from a significantly lower bandwidth recording at a later stage. The former requirement can be further refined into real-time recovery requirements; in this case the watermark must be recovered given a signal excerpt with a duration delimited by a fixed upper time bound and given a fixed upper bound for the time permitted to recover the watermark after the signal excerpt has been available.

For digitally transmitted signals, it must not be possible to detect (and therefore delete) the marking without an appropriately parameterized detector from either the encoded or the baseband (decoded) signal and must be robust against digital-to-analog conversions. Since most multimedia signals transmitted digitally are encoded using a compression scheme and have only a fixed bandwidth available, an additional requirement levied on digital watermarks may be that the watermark does not increase the bandwidth required for the marked signal beyond the available bandwidth for a given signal.

3.2 Copyright protection

The protection of intellectual property through technical means was presumably one of the primary motivations for applying well-known steganographic techniques; some of the earliest publications in the field explicitly mention this particular application area [1, 2].

Protective measures can be grouped into two broad categories. The first category encompasses the protection against misappropriation of creations by other content providers without the permission of or compensation of the rights owner, while the second category includes protection mechanisms against illicit use by end users.

While both categories are commonly referred to as piracy, the issues involved and requirements for protective measures may make a distinction beneficial, although in a number of application scenarios protection against both categories of misuse are called for.

In addition to the partial requirements derived for the protection scenarios given below, there exists an orthogonal requirement component in the quality of the signal to which digital watermarks have been applied; presumably the most common quality metric is the absence of perceptually significant artifacts introduced by markings. Such a quality metric must be based on an accurate perceptual model; even though simpler metrics such as peak signal-to-noise ratio (PSNR) can provide guidance in the application of algorithms and their parameters, effects such as masking in the audio and color sensitivity in the visual domain cannot be adequately covered.

However, while human test subjects potentially provide the most elaborate evaluation metric, the cost of deriving statistically significant data eliminating personal preferences and bias makes the approach impractical under most circumstances. Instead, computational models derived from such experiments can be employed that can predict the effect of a modification to a signal on the perceived quality (e.g., a just noticeable difference). The accuracy and level of detail of the perceptual model used for evaluation therefore are critical; however, given the derivation mechanism for the models, the ability to, for example, state that a specific individual will or will not perceive an alteration for a specific signal after the application of a marking is limited to probability statements over populations.

The notion of perceptual significance is not necessarily limited to the human audiovisual system; in some scenarios, markings may identify ownership and distribution of multimedia data that are not intended for human perception, but rather for machine analysis. In this case, potential artifacts introduced by the marking process must not affect the acquisition and registration process of the automated analysis or be interpretable as semantically significant. However, this scenario is considerably more benign compared to one targeted to the human perceptual system, as the model is well defined and can be verified. Additional nonperceptual quality metrics (subsidiary to the ones discussed above) may include the effect a marking has on the compressibility of the signal. Any copyright protection system must ensure that the false positive rate (depending on the application subscenario, this can mean that the probability that a digital watermark detector responds without a marking being present or that an existing marking is read, but the payload is not retrieved correctly in such a way that another syntactically correct payload is detected) is below a certain threshold, transgression of which would lead to increased expenses due to litigation, warranty claims, or other customer dissatisfaction. In a consumer-oriented application area, this imposes a significant burden of proof on the digital watermarking algorithm (e.g., the digital versatile disk (DVD) Copy Control Association (CCA) requires a false positive rate below 10^{-12} [3]).

3.2.1 Misappropriation by other content providers

Misappropriation by other content providers can, in turn, occur in several forms. In the simplest case, a creation is duplicated, redistributed, or resold in its entirety and in its original form. Here, the legal framework discussed in Chapter 1 provides, in its current form, protection even without the creation being marked with a copyright notice, although certain national jurisdictions may provide elevated protection status for creations affixed with a formal copyright notice. For most types of creations in digital representation such as movies, books, or even software, the addition of such a notice does not impose any difficulty and can be embedded in textual form. Other media types, particularly audio material, cannot be annotated in such a simple way. In addition, visible markings may not be desirable because of a concomitant loss of value to the end user; such a perceptible marking can be distracting or esthetically displeasing and would have to be placed in a semantically or esthetically significant location, since otherwise a simple cropping operation would remove the marking.

Digital watermarking, or indeed a steganographic marking, can provide the requisite embedded information to assert copyright over a creation. In the most trivial form this amounts to embedding a source-encoded textual notice, but given the limited bandwidth afforded particularly by robust digital watermarks, more efficient encoding schemes are called for. Generally, this application scenario requires that information serving as evidence of the origin of a creation is provided as the payload signal.

A more realistic application scenario must recognize that the creation may be subjected to a number of transformations including format conversions. Even for the case of duplication of the entire creation without alteration, a conversion in the representation may remove copyright notices that are not bound to the actual carrier signal of the creation, again making digital watermarks and steganography attractive.

Many transformations to which a creation may be exposed either deliberately or unwittingly are, however, not lossless. Such transformations may include the use of lossy compression to reduce storage and bandwidth requirements, digital-to-analog conversions, or complex operations such as multiband compression in the case of audio data that introduce a significant distortion compared to the original creation. It is primarily due to these considerations that the robustness requirement of digital watermarks for copyright protection was introduced.

Unauthorized duplication can not only occur in situations with no preexisting relation between the rights owner and the secondary user, there are also situations where a user has licensed a creation for limited exploitation under certain circumscribed conditions. Any usage outside the licensed area must be unattractive to the secondary user, which can be accomplished by means of digital watermarking [4] in that the digital watermark contains evidence of the origin of the material. This immediately leads to an additional requirement for the application scenario: there must exist a surjective mapping between the set of rights owners O and protected creations C (in most situations there will be additional requirements on the relation between the sets, but for the purposes of this discussion, surjectivity is sufficient to capture the general case).

Establishing such a mapping using digital watermarks implies that the payload size for the ownership marking must be sufficiently large to accommodate the domain of the mapping.

The requirement can, however, also be fulfilled by a marking that permits the explicit identification of each element of C, in effect providing an efficient annotation watermark (see Section 3.3). In this case, however, the payload size may be significantly larger than in the previously mentioned one.

An additional payload type, identifying individual copies of a creation as associated with a specific transaction, can also provide a deterrent effect; this application scenario for digital watermarks is discussed in Section 3.2.2.

While, as noted above, the creation may be subject to quality-lowering transformations, the application scenario dictates that the secondary user cannot degrade the quality to a significant extent, as that would likely reduce the value perceived by the secondary user's customers to an extent that makes the duplication unattractive, in contrast to the requirements derived in Section 3.2.2, although the quality degradation and variety of transformations inherent in a digital-to-analog conversion may be acceptable, as made evident by [5]. This particular observation is rather problematic for protection mechanisms based solely in the digital domain, as discussed in Chapter 8. Robustness requirements, particularly with regard to lossy compression, are therefore modest if the end result after an attack is to be of value to a potential adversary. However, a high degree of robustness against desynchronization attacks is desirable, as such an attack does not necessarily degrade the perceived quality (e.g., shifts of single scan lines in a motion picture).

A significant threat related to desynchronization is inherent in the nature of digital media in that the composition of new creations using parts of one or more existing ones can be performed with comparable ease and modest tools.

Such compositions can take the form of collections of existing creations, in which case the precise technical form of the collection (such as the use of hyperlinked material instead of locating material at the same source) can determine the admissibility of collections without requiring compensation for the rights owner of the component creations, depending on the doctrine in a given jurisdiction.

Another type of composition, which may even occur without a deliberate attempt at deception on the part of the secondary user is in the composition of fragments of existing creations, which may occur over multiple generations. If only one of the intermediate generations omits the required copyright notice, it becomes increasingly difficult to associate the fragment with its rights owner. While such composition may be legitimate in the case of very limited fragments under the fair use doctrine, the compensation of the rights owners cannot be avoided in general. In addition, the rights owner may object to having a creation used in a certain context and refuse to grant permission for such uses (e.g., the combination of unrelated video material designed to create a misleading impression; this is closely related to the problem described in Section 3.6).

The difficulty with which such fragments could ordinarily be traced back to their respective origin not only endangers the rights of creations' owners, it also imposes a severe burden on conscientious secondary user attempting to identify and then locate each creation's owner for obtaining permission. A digital watermark identifying the creator directly or indirectly is therefore beneficial not only to the rights owner but also to legitimate secondary users, as it can significantly reduce the effort required and may make certain kinds of compositions possible that could otherwise not have been considered.

Besides introducing a requirement for robustness against trivial desynchronization, such applications add further constraints on the capabilities required for watermarking techniques in that the watermark must be recoverable after extensive cropping. Depending on the type of media, this may, for example, involve spatial cropping in the case of video or still images, or temporal cropping for audio and video material. However, the watermark payload must still be retrievable from the cropped elements, for example, by having the signal spread out with sufficient redundancy or repeating the watermark over the course of the carrier signal. This implies that the bandwidth requirement for the watermark payload applies not to the entire signal but rather to the minimum fragment size for which recognition must be possible. In case of audio data, the fragment size is between 3 and 5 seconds; for other media types, the minimum fragment size may depend on the semantics of the fragment, for example, in the case of a visible trademark or trade secret stemming from a reproduced document, the same signal fragment extent is significant, whereas the recognition of the origin of a semantically irrelevant fragment of the same size may not be required.

3.2.2 Illicit use by end users

The main distinction between the unauthorized reproduction and use of creations by content providers discussed in the previous section and unauthorized use by end users lies in the visibility of the perpetrator. Whereas a content provider ultimately must attempt to sell or otherwise profit from the creations and therefore must expose himself to the public (albeit possibly in a different jurisdiction), the same is not true for end users.

End users may duplicate creations and distribute such material either within their personal environment or using file sharing services, some of which can provide a certain amount of anonymity. While copyright markings and digital watermarks such as those discussed in Section 3.2.1 can assist in identifying such material if and when it is located, identification of the source of the material cannot be accomplished using these markings.

A deterrence effect may, however, be conjectured if an individual copy of a creation is tied to a specific transaction (which may implicitly be extended to an individual based on the type of records maintained for a transaction). The payload for such a digital watermark may be the identity of the end user to which a creation is sold or otherwise licensed or a unique identification of the transaction itself.

This can lead to the identification of the original purchaser or licensee (or the last authorized link in a distribution chain in the case of what has occasionally been called *superdistribution*) if a copy or elements thereof are found in the possession of an unauthorized end user.

The payload size required for this application scenario mirrors those discussed in the preceding section; for transaction watermarks, the uniqueness constraint must be balanced against the drawbacks of large payloads, at least to an extent that the probability of duplicate transaction identifiers is comparable to other types of false positives in the detection stage, since otherwise the evidentiary value could be called into question.

Pragmatic issues also must be taken into consideration in determining the true deterrence effect of marked creations, since—unlike the case of commercial interests—in most situations there will exist a strong legal protection of individuals from searches of their property and invasion of their privacy without a viable justification. This is likely to limit the deterrent value in that only copies that are found in the open (e.g., those traded openly by an end user) can be verified for containing watermarks. It is also, to the authors' best knowledge, untested whether such prima facie evidence in the form of a digital watermark is sufficient to show that a transgression has taken place—and even so it may be the case that an end user thus identified can plausibly deny the deliberate dissemination.

An argument for the efficacy of digital watermarks as a deterrent in this context is the use of automatic search engines that scan for protected creations. Such searches can occur either by transferring suspicious content to a central location and analyzing the data there or by using so-called agents to have the analysis process take place in situ [6].

In the case of the application scenario discussed in the preceding section, central analysis is already exceedingly difficult from a logistical standpoint (additional technical aspects are discussed in Chapter 7), as the set of potential sources for redistribution is not properly bounded and may grow faster than the product of bandwidth and processing power available to the rights owner. The approach becomes even less attractive when arbitrary end user systems are considered—these will generally not make creations available to external nodes and also may not be available at the time of checking for protected creations [7].

Using agents to detect misuse could, under very benign circumstances, eliminate the logistical problem in the scenario considering misappropriation by other content providers—although making processing power available to rights owners through content providers would almost certainly require commensurate reimbursement, but would leave the problem of the ill-defined set unsolved. It is, however, unlikely that any end user would consent to rights owners executing arbitrary code and granting access to any data located on the end user's system—even though a comparable approach has been proposed in a similar context for a digital rights management system [8].

The robustness requirements for digital watermarks protecting against end users are considerably higher, since the quality aspect appears to be of lesser significance if creations are obtained for free or for the cost of transferring the data. This assertion is supported by the observed popularity of highly compressed representations of audio and particularly video data, the latter at a quality that is significantly below the level of the original.

Given such low-quality requirements, robustness must also be maintained against a number of deliberate attacks (as discussed in Chapter 7). This is particularly problematic, since there exist automatic tools for performing—typically highly successful—attacks against digital watermarking systems (e.g., used in developing benchmarks or in the course of academic research) that can be used even by individuals with modest skills which introduce quality degradations comparable to or in most cases significantly less than those tolerated by end users in the case of compression [9].

3.3 Annotation watermarking

Digital watermarks possess a number of desirable properties that make their use outside of copyright protection desirable, some of which are analogous to the ones discussed in the preceding sections.

Annotation watermarks [10] provide information in a side channel that is coupled to the carrier signal without degrading the perceived quality of the carrier signal; this distinguishes them from markings that are either visible or not tied to the carrier signal (e.g., comment fields specific to a certain file format). They are therefore of interest in applications where the format of multimedia data cannot be guaranteed or is likely to change throughout a work flow; similarly, if digital-to-analog conversions are part of the expected transformations that multimedia data must resist, the properties of digital watermarks are desirable.

Unlike perceptible markings, digital watermarks can also be distributed across an entire carrier signal (e.g., a video stream) such that the resulting signal can be cropped significantly and the digital watermark can still be recovered either in its entirety or to a significant extent.

It should, however, be noted that in some applications, such as those involving images or video data, watermarks are appropriate only if the requirements for constant quality or robustness of the marking to cropping are given. Otherwise, there exist alternative technologies (e.g., two-dimensional bar codes) that can provide considerable payload capacity with comparable or better robustness characteristics.

The annotation application scenario can be considered notably distinct from steganographic techniques in that the presence of the watermark signal is public knowledge (a property that may, for example, also be true for copyright protection watermarks if they are employed as a deterrent) and a detector may also be available to the public.

Unlike, for example, a copyright protection scenario, no immediate adversarial relation needs to be stipulated, since the marking itself is not directed against the interests of a particular individual or group.

Given the robustness of most digital watermarks particularly as applied to digital-to-analog conversions, one of the prime uses of annotation watermarks is the association of an analog representation with its digital original; this can, for example, occur via centralized database records, limiting the payload requirement to a single unique key for such a database. The ability to reference the original, given possibly only a cropped or otherwise partial copy of a document or other multimedia data, significantly eases record handling and can enable multimedia document management systems. An example of such an application scenario would be the use of X-ray photography for testing and documenting the structural integrity of materials and construction. Here, excerpts from the original (possibly very large) analog photograph can be scanned and reproduced. It is, however, important to retain information as to the precise circumstances under which the photograph was originally taken and, from there, to any related documents. Somewhat problematic in this particular application scenario (and even more so in the case of medical

imaging, where—at least in principle—similar concerns exist) is the requirement for very high fidelity of the marked representation and the implicit
requirement that no artifact introduced by the watermarking process may
resemble a feature that can be misinterpreted by an analyst.

Regardless of whether the payload for the digital watermark is the annotation itself or a key into a database containing the actual referenced records,
safeguards such as error-detecting or error-correcting codes must be employed to protect the integrity of the annotation. Conversely, some applications for annotation require that the integrity of the annotated signal be
preserved. This can take the form of several possible subrequirements; in a
simple case, the duplication and transfer of an annotation watermark without
authorization to another carrier signal must be prohibited. A more elaborate
requirement is that the semantic integrity of an annotation-marked signal
must be preserved. This requirement can, to some extent, be satisfied by
the use of fragile watermarks, but—as noted in Chapter 2—this implicitly
contradicts several robustness requirements.

The application scenarios described in Sections 3.4 and 3.5 can be considered a specific subscenario of annotation watermarking.

3.4 Fingerprinting

The figurative term *fingerprinting* has acquired two completely disjointed interpretations in the field of content protection, only one of which applies
to digital watermarking (a third interpretation involves the derivation of a
unique characteristic of a creation with significant, typically constant, bandwidth such as a cryptographic hash function; as this derived information is
disjoint from the creation or carrier signal, this type of fingerprinting is of no
particular interest here by itself).

In the watermark interpretation (which semantically predates others, the
general terminology having been introduced by Wagner [11] even prior to
the application to digital watermarks), a digital watermark uniquely identifying the end user of a creation is embedded in the creation's carrier signal as the payload, corresponding to an application subscenario described in
Section 3.2.2. This implies the same requirements for the payload size of the
watermark and the very high requirements for robustness against deliberate attacks specific to copyright protection scenarios. The fingerprint watermarks can be embedded at the time of distribution to a specific customer;
this requires a considerable computational overhead for the generation of
watermarks as well as a distribution medium that permits the efficient creation of distinct copies of a creation. Alternatively, a playback device that

contains a subsystem tied to a specific individual or customer can embed a fingerprint watermark immediately on playback. The latter approach does not require distinct copies and reduces the computational burden on the content provider. This scenario was used by the now-defunct DiVX pay-per-view digital video player scheme developed and owned by a partnership between Ziffren, Brittenham, Banca & Fischer, an entertainment law firm in Los Angeles, California, and Circuit City.

However, fingerprint marking on playback implies that both the embedding mechanism and the requisite key material are present in the playback device (even download of ephemeral key material does not alter this situation) and hence under the control of a potential adversary. Implicit in this observation is the need for a separate embedding key for each playback device, since otherwise any single compromised playback device would mean that the adversary can embed arbitrary fingerprints, eliminating the evidentiary value of the watermarks.

However, this implicit requirement for key material imposes a severe computational burden in case a fingerprint needs to be identified, since for each suspect device fingerprint, a test for the existence of a watermark must be performed.

The other interpretation refers to the extraction of semantically relevant or characteristic features from multimedia signals to identify the signal itself and does not involve digital watermarking at all. This technique is described in Section 9.1.

3.5 Automatic playlist generation for rights verification

Broadcasts or other types of public performances of creations must be accompanied by appropriate compensation for the rights owners. While there exist societies offering centralized records keeping and compensation of individual rights owners in most jurisdictions (e.g., ASCAP for music in the United States), it is nonetheless burdensome and error-prone work to create the requisite records on which creation was played, when it was played, and how many times.

Inserting an annotation watermark into each individual creation permits the automatic monitoring of a broadcast stream or similar performance. This lowers the reporting burden on the performing entity and can therefore be performed at the originating site, resulting in lowered robustness requirements due to a lower risk of distortions introduced by compression, transmission faults, and incomplete reconstruction. In such an application subscenario, the bandwidth (typically time) required for a given digital watermark is of secondary interest.

However, a secondary benefit to royalty-collecting entities can be derived only if playlist monitoring occurs after performance or broadcasting.

In this case, the time required for the recovery of a digital watermark is critical to the overall efficiency of the monitoring scheme, since if each marking requires only a fraction of the duration of a creation, multiple signal sources can be monitored simultaneously by moving to a different source once a marking has been detected. Assuming that such a scan cycle does not last more than the average length of a creation and a low unit value for individual royalty payments, this reduces the expenditures necessary for the monitoring equipment and bandwidth. Even more cursory monitoring would also be adequate for this application scenario, since typically only gross or systematic underreporting is of actual interest.

Given that almost all creations require the payment of royalties for performance, the threat of deliberate removal on the part of the entity broadcasting or performing the creations would be limited, as the lack of annotation watermarks would be sufficiently abnormal to warrant a manual inspection of the material, presuming that the presence of annotation watermarks was mandated by the royalty-collecting entity.

The automatic generation of playlists is also relevant to the reverse case, that is, to verify the fact that a given creation (advertisement) has been broadcast according to a previously established contract. As in the previous case, the robustness requirements are derived mainly from the need to withstand standard processing chains encountered, so deliberate attacks on the watermark are of limited utility to the parties involved.

3.6 Multimedia authentication

As noted in Section 1.3, the integrity and authenticity of multimedia signals, particularly those already in digital representation, are in jeopardy of malicious or otherwise semantically distorting manipulation from their creation or reception by a sensor onward.

For application areas where the bitwise identity and authenticity of a digital document is either a priori desirable (e.g., in the case of documents in electronic representation) or otherwise feasible, cryptographic hash functions and digital signatures (see Section 2.2) can provide both effective and efficient protection.

However, for most applications involving multimedia signals, certain types of modifications such as compression, resulting in a bitwise difference between signals, will still result in a signal that is considered authentic and for which integrity is not considered violated.

Given an original or authentic digital multimedia signal S and a signal S', which is a putatively transformed S, several problems can be formulated. The authenticity of S' can be shown with or without knowledge of the original S. In the first case, this is trivially accomplished by comparing the signals under a proximity metric. In the second case, however, the authenticity must be determined based on a characteristic feature. Such features—if the requirement for similarity is maintained; otherwise solutions such as Friedman's trustworthy digital camera using a signed hash value transmitted out of band [12] would be sufficient—should be intrinsic to the signal to avoid loss of the feature due to legitimate processing resulting in a loss of integrity and authenticity information while the signal itself is still valid.

Digital watermarks provide an obvious solution to some of the requirements described above. However, the very robustness of regular digital watermarks against manipulation (especially against cropping and spatially or temporally localized alterations) makes these markings unsuitable. Instead, fragile watermarks (see Section 2.3.2) are called for [13–15] particularly in the case of integrity protection applications. The drawbacks inherent in first generation fragile watermarks imply that a number of processing steps will result in the watermark being unrecoverable even though the integrity criteria could still be met for a given S'. The main problem in this case is the formulation of a similarity metric that determines the semantic equivalence (for a given application) of two signals S, S' and subsequently the construction of a detector that signals the recovery of the fragile watermark if a given S' exceeds a similarity threshold.

Using watermarks for authenticity, and to some extent also for integrity, imposes the strong requirement (see Chapter 7) that no unauthorized entity can embed a marking that purports to be another entity, and that the marking is linked in such a way to the carrier signal that a transfer of the marking from one carrier signal to another carrier signal resulting in detection of the authentication feature is not possible.

3.7 Watermarking for copy protection

Digital watermarks can be considered protection techniques only in that they provide a deterrence mechanism or evidence of breach of copyright or other contractual obligations after the fact. There exist, however, approaches for utilizing watermarks immediately for copy protection in conjunction with specially equipped devices for playback and recording.

Protection can occur at two separate stages, during recording and playback. In each case both the presence of a potentially specific marking or the absence thereof can be used to induce a desired behavior of the device controlling the operation.

Requiring the presence of a watermark to permit playback of a creation bears some resemblance to the authentication scenario discussed in Section 3.6; for digital media representations, this application scenario implicitly requires the personalization of the creation for a given individual or set of devices, since otherwise a successful duplication of the digital representation would also reproduce the watermark; depending on the robustness requirements for the watermark, watermark recognition may even be possible from copies generated from analog sources (e.g., audio signals captured and redigitized from an analog output of a legitimate playback device).

The requirements for payload capacity match those discussed in Section 3.2.2 in case users (or devices) are identified by the transaction watermark; for individual transaction records, the payload size is correspondingly higher. As copies must be individually marked in this application scenario, this imposes limitations on the distribution forms that can be justified economically.

Requiring the absence of a watermark for playback could, for example, be part of an application scenario in which a playback device embeds a watermark (such as its identity) into the creation as it is played back—alternatively this watermark could also be embedded in a recording process—to identify first from subsequent generations of copies. This is the application scenario most similar to the *copy bit* approach found in digital audio tape (DAT) and audio CD systems, along with the familiar threats and vulnerabilities from that approach—albeit with an increase in difficulty if the embedding process is an integral part of decoding a creation for playback.

Conversely, another application scenario consists in requiring the absence of a digital watermark for recording. While such a scheme is only of interest in cases where it can be guaranteed that any recording device honors this convention, the benefit compared to a simple copy bit mechanism is that removal of the marking once it has occurred requires more effort than would be the case of a marking that is not tied to the content itself.

In addition, more elaborate schemes can use the fact that payload sizes larger than a single bit can encode (assuming safeguards against unauthorized manipulation of the payload or the marking itself) arbitrary instructions as to the admissibility of copying or playback operations that can be changed dynamically either in the case of duplication or—provided a writable representation—during the use of the representation, for example, to record the number of remaining playback operations for a given license. As noted above, such a mark-on-use scheme was used for the output of the DiVX devices for the playback of digital movies, although the digital watermarks were embedded only in the analog signal and not used for copy protection as such.

References

[1] Cox, I. J., et al., *Secure Spread Spectrum Watermarking for Multimedia*, Technical Report 95-10, NEC Research Institute, 1995.

[2] Koch, E., and J. Zhao, "Towards Robust and Hidden Image Copyright Labeling," in I. Pitas, (ed.), *Proceedings of 1995 IEEE Workshop on Nonlinear Signal and Image Processing*, Neos Marmaras, Greece, June 1995, pp. 452–455.

[3] DVD Copy Control Association, *Request for Expressions of Interest*, technical report, DVD Copy Control Association, April 2001.

[4] Roth, V., "Sichere verteilte Indexierung und Suche von Digitalen Bildern," Ph.D. thesis, Darmstadt Technical University, Darmstadt, Germany, 2001.

[5] International Federation of the Phonographic Industry, *IFPI Music Piracy Report*, June 2001.

[6] Zhao, J., and C. Luo, "Digital Watermark Mobile Agents," *Proceedings of NISSC'99*, Arlington, VA, October 1999, pp. 138–146.

[7] Perrig, A., and A. Willmott, *Digital Image Watermarking in the "Real World,"* technical report, Carnegie Mellon University Computer Science Department, Pittsburgh, PA, January 1998.

[8] Levy, S., "The Big Secret," *Newsweek*, July 2002.

[9] Craver, S. A., et al., "Reading Between the Lines: Lessons from the SDMI Challenge," in *Proceedings of the 10th USENIX Security Symposium*, Washington D.C., August 2001.

[10] Bender, W., et al., "Techniques for Data Hiding," *IBM Systems Journal*, Vol. 35, Nos. 3 & 4, 1996, pp. 313–336.

[11] Wagner, N. R., "Fingerprinting," *Proceedings of the 1982 IEEE Symposium on Security and Privacy (SOSP '83)*, Oakland, CA, April 1983, pp. 18–22.

[12] Friedman, G. L., "The Trustworthy Digital Camera: Restoring Credibility to the Photographic Image," *IEEE Transactions on Consumer Electronics*, Vol. 39, No. 4, November 1993, pp. 905–910.

[13] Walton, S., "Image Authentication for a Slippery New Age," *Dr. Dobb's Journal of Software Tools*, Vol. 20, No. 4, April 1995, pp. 18–20, 22, 24, 26, 82, 84–87.

[14] Yeung, M. M., and F. C. Mintzer, "Invisible Watermarking for Image Verification, *Journal of Electronic Imaging*, Vol. 7, No. 3, July 1998, pp. 578–591.

[15] Kundur, D., and D. Hatzinakos, "Digital Watermarking for Telltale Tamper Proofing and Authentication," *Proceedings of the IEEE*, Vol. 87, No. 7, July 1999, pp. 1167–1180.

CHAPTER

4

Contents

Digital watermarking for still images

A lot of research effort was spent on the development of watermarking algorithms for images, which is discussed in this chapter. Starting with a short summary of application scenarios of image watermarking techniques, the evolution of image watermarking techniques for photographic images is outlined. This is followed by a section dealing with the watermarking principles for binary and halftone images. A short summary finalizes this chapter.

4.1 Classification and application requirements

Watermarking techniques are applied to images because of various reasons. Each of these possible applications involves typical processing operations that a watermarking technique must survive. *Content protection* scenarios may include operations like color to gray-scale conversion, global or local affine transforms, and printing and scanning. *Authentication* watermarks must not be affected by legal operations, while illegal attacks must destroy them. Metadata labeling scenarios may include media transform. A typical example is the transmission of information in printed images. This information is revealed if the printed image is shown to a webcam whose data is processed with the watermark reader software as presented by Digimarc [1]. Yet robustness is not a general requirement for *data hiding* techniques: Undetectibility is essential. A typical scenario for data hiding is the distribution of hidden information via (Usenet) newsgroups, bulletin boards, or simply by images on homepages. Steganalysis is a new research area dealing with the detection of hidden data as presented, for example, by Fridrich and Goljan [2]. A possible application of these techniques is the so-called StegoWall as proposed by Voloshynovskiy et al. [3]. This StegoWall can be compared with a firewall that analyzes the data that should be transmitted and prevents the transmission of any data containing hidden information. For additional discussion of applications, see Chapter 3.

4.2   Photographic and photorealistic images

In general, most of the watermarking methods described in this section can be applied to color as well as to gray-scale images if the embedding of the watermark is not directly dependent on the color information of the image. Therefore, embedding in the intensity values is proposed in numerous image watermarking publications.

The first method proposes embedding the watermark information in the LSB [4]. The principles of direct LSB encoding are given in Chapter 2. The advantage of the direct LSB method is its capacity. A color image with the typical size of $1,600 \times 1,200$ in red, green, blue (RGB) representation is capable of storing more than 700 kB even if only the LSB is used as information carrier. If visible artifacts can be ignored, more bits can be used for hiding information. The main disadvantages of this simple approach are its lack of robustness against lossy compression, its visible artifacts especially in flat image regions, and statistical dependencies that can be detected. More sophisticated methods are presented, for example, by Westfeld [5], Lee and Chen [6], or Crandall [7]: LSB techniques are not limited to the spatial domain. They can also be applied to the image representations in transform domains. Embedding in a transform domain can be used to change the statistical behavior of the information carrier. Additional improvement can be achieved by carefully choosing the embedding scheme. The missing robustness and fragility of the LSB method is not a general disadvantage. In certain application scenarios, for example, for image authentication, fragility is a desirable criterion.

4.2.1   Traditional watermarking methods

The abstract model for watermarking is independent of the underlying media type and is shown in Figure 2.5. But embedding and detection, especially the resulting artifacts, are dependent on the data type. Traditional image watermarking techniques are based on spread-spectrum communication, as described in Section 2.3.3. Moreover, frequencies carrying additional information cannot be modified arbitrarily because of the concomitant image degradations. Modifications of low frequencies affect the mean intensity and result in the noise of low spatial frequencies. These effects are strongly visible.

On the other hand, modifications of high frequencies result in less visually distracting high-frequency noise; more details about the perception of distortions in images are given in Section 7.3.3. The application of suitable image filters, which are commonly available in image processing programs (e.g., for image enhancement and intrinsically in image compression algorithms), can remove data embedded in high frequencies. Medium frequencies can generally be considered suitable information carriers to fulfill the requirements of robustness as well as of low visible degradation. Models of human perception (see Section 7.3.3) are applied for increased performance.

"A Digital Watermark" is the title of the paper published by van Schyndel et al. [4] in 1994. They propose two methods for watermarking digital data. Both methods use m-sequences to derive a pseudonoise (PN) code (the watermark). The first method compresses the image data from 8-bit gray scale to 7 bits by adaptive histogram modifications. The LSB is directly used to embed the watermark information. The second proposed method uses LSB addition: The watermark is added to the LSB plane. For retrieving the watermark, a correlation-based extraction scheme is used as discussed in Section 2.3.3.

Parallels between spread-spectrum communications and watermarking are first considered and discussed by Cox et al. [8, 9]. This algorithm, whose scheme is shown in Figure 4.1, uses a frequency domain transform to convert the input image into another domain.

In the frequency domain, a sequence of values $c_o = c_o[1], \ldots, c_o[n]$ is extracted from the image. This sequence is the information carrier of the watermark and is modified. The watermark is a sequence of real numbers $w = w[1], \ldots, w[n]$. Each value $w[i]$ is chosen independently according to $N(0, 1)$ (Gaussian distribution with mean $\mu = 0$ and variance $\sigma^2 = 1$).

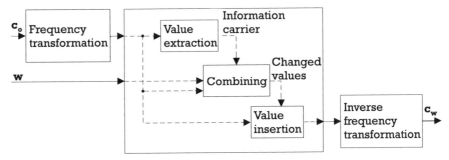

Figure 4.1 The embedding scheme of the watermarking algorithm as proposed by Cox et al. [9]. A lot of the proposed watermarking techniques share the same model.

Three different formulas are suggested by the authors for embedding, whose difference lies in their embedding characteristic and in their invertibility:

$$c_w[i] = c[i] + \alpha w[i] \qquad (4.1a)$$

$$c_w[i] = c[i](1 + \alpha w[i]) \qquad (4.1b)$$

$$c_w[i] = c[i] \exp(\alpha w[i]) \qquad (4.1c)$$

The scaling or watermark strength parameter α influences the robustness as well as the visibility of the embedded watermark. In the previous equations, α is a global scaling value. Better results can be achieved by using multiple (e.g., image region–dependent) scaling parameters. Details on how to determine the scaling parameter α and the fact that under some assumptions these three equations can be considered as equal are described in [10].

In their original publication, Cox et. al suggested using the discrete cosine transform (DCT) domain, although other transform domains are applicable, too. The result of the insertion of the combined values c_w into the original image is the watermarked image. Watermark retrieval is also based on correlation.

As mentioned before, embedding a watermark by this method is not limited to the DCT domain. However, the DCT domain has been extensively studied because this is the transform used in Joint Picture Expert Group (JPEG) compression, where extensive studies on perceptibility were performed. Further advantages of using the DCT domain include the fact that frequency decomposition in frequency bands is efficient, DCT transform is widely used in image and video compression schemes, and the DCT coefficients affected by compression are well known.

A considerable number of image watermarking techniques share this architecture. Yet they differ chiefly in the signal design, the embedding, and the retrieval of the watermark content [11]. In the following paragraphs we will give a short summary of the different embedding and retrieval methods.

The DCT transform is a special case of the discrete Fourier transform (DFT). The DFT can be described as

$$F(k_1, k_2) = \frac{1}{N_1 N_2} \sum_{n_1=0}^{N_1-1} \sum_{n_2=0}^{N_2-1} f(n_1, n_2) \, \exp\left(-i2\pi \frac{n_1 k_1}{N_1} - i2\pi \frac{n_2 k_2}{N_2} \right) \qquad (4.2)$$

Amplitude modulation of the DFT coefficients is applied by many watermarking techniques [12]. One advantage of the DFT transform is the resulting shift (translation) invariance. Another one is the ease of considering the human perception by weighting frequencies. The properties of the DFT have been studied extensively in image processing literature. One of the results obtained there is the fact that the phase information is more important for the image content than the magnitude [13]. Consequently, Ó Ruanaidh et al. [14] propose embedding the watermark in the phase information of the image, which is comparable to phase modulation in communication theory, in contrast to the previously described amplitude modulation. For a blind retrieval of the watermark, an optimal statistical detector is proposed by Ó Ruanaidh et al.

Various methods for watermarking digital images in the wavelet domain have been proposed. Among other reasons, the development of new compression schemes led to new watermarking techniques.

Barni et al. proposed a watermarking method based on the wavelet decomposition [15]. The wavelet decomposition decomposes the input image in high and lowpass components with different orientations [16]. A three-level decomposition of an image is shown in Figure 4.2.

For watermarking, a discrete wavelet transform (DWT) is applied to the original image. The watermark is inserted in the highest level detail subbands according to the following rule:

$$c_{\mathbf{w}}^{LH}[i, j] = c_{\mathbf{o}}^{LH}[i, j] + \alpha \lambda^{LH}[i, j]\mathbf{w}[iN + j] \tag{4.3a}$$

$$c_{\mathbf{w}}^{HL}[i, j] = c_{\mathbf{o}}^{HL}[i, j] + \alpha \lambda^{HL}[i, j]\mathbf{w}[MN + iN + j] \tag{4.3b}$$

$$c_{\mathbf{w}}^{HH}[i, j] = c_{\mathbf{o}}^{HH}[i, j] + \alpha \lambda^{HH}[i, j]\mathbf{w}[2MN + iN + j] \tag{4.3c}$$

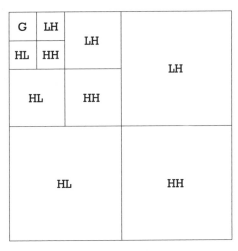

Figure 4.2 The wavelet pyramid of a three-level image decomposition. Each level consists of three detail images, which correspond to horizontal, vertical, and diagonal details of each resolution level. *xy* stands for applying filter *x* in row direction and filter *y* in column direction, where *L* means lowpass filtering, *H* highpass filtering, and *G* is the coarse image.

where α is the global parameter for the watermarking strength, λ is a local weighting factor and $\mathbf{w}$ is a pseudorandom binary sequence. This local weighting factor considers the masking characteristics of the human visual system (HVS; see Section 7.3.3). For retrieving the watermark, the correlation between the watermarked DWT coefficients and the watermarking sequence is computed. The authors also propose a method for choosing the threshold in a way to minimize the false positive detection probability. Similar techniques have been proposed by other authors, as in [17, 18]. A survey of different wavelet-domain watermarking algorithms is given by Meerwald and Uhl [19].

The patchwork technique (see Chapter 2) is also applied to image watermarking, as, for example, proposed by Pitas and Kaskalis [20]. They split the image into two subsets, and in one subset the pixel values are increased, whereas in the other subset pixel values are decreased. Further patchwork techniques are block based, like the techniques proposed by Langelaar et al. [21] or Bruyndonckx et al. [22].

A JPEG model–based watermarking method as presented by Burgett et al. [23] was the first efficient watermarking in the DCT domain [11]. The design of this method is patterned along the lines of the JPEG compression model. The image is divided into square blocks. The size is chosen according to the JPEG compression: 8×8 pixels. Each block is converted into the DCT domain. The DCT blocks are chosen pseudorandomly for embedding the watermark. The watermark is embedded in each block by changing selected pairs of coefficients. Figure 4.3 shows the coefficients from which pairs are

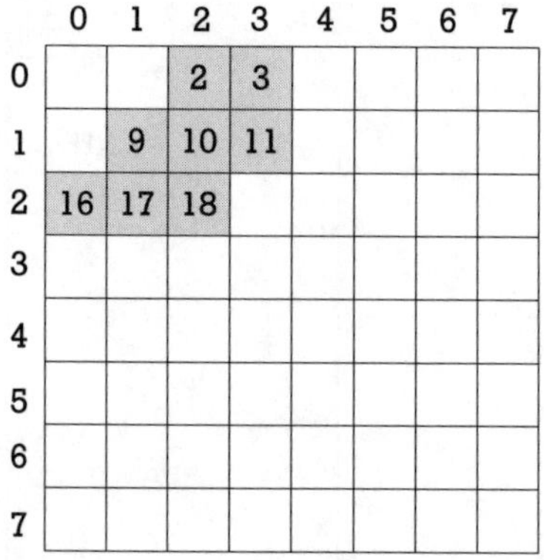

Figure 4.3 The coefficients of the DCT blocks used for embedding the watermark by Burgett et al. [23]. These coefficients represent the middle frequencies of the image. Information is embedded by changing the relation in a pair of coefficients.

chosen according to a key and according to the information of the frequency band in which the information should be stored. The embedding information determines the resulting difference, where its sign is used as information carrier. For improved robustness against JPEG compression, the quantization matrix is considered to adjust the strength of the watermark. During the watermark retrieval process, the differences of the coefficient pairs have to be evaluated.

The similarity between this method and the correlation-based method is shown by Cox et al. [10]: This algorithm can be formulated as a correlation by defining a pattern with the same dimensions as the coefficient matrix. The pattern values are determined by the influence of the corresponding coefficients: It is zero for coefficients not considered in the evaluation. The pattern values for the pair coefficients are either 1 or −1. Thus, the sign of the correlation directly depends on the relation of the pair coefficients.

Watermarking algorithms based on quantization are proposed by Chen and Wornell [24] and described in Section 2.3.3. Another evaluation of a quantization-based watermarking scheme is given by Eggers and Girod [25]: The authors propose the embedding of a semifragile watermark for content authentication. The method proposed by Kundur and Hatzinakos [26] is based on the wavelet decomposition or wavelet pyramid. The first stage includes the computation of the Lth-level discrete wavelet decomposition. Each level consists of three detail images, which correspond to horizontal, vertical, and diagonal details of the L resolution level. Each detail image can be represented as $\mathbf{c}_{\mathbf{o}k,l}[m, n]$, where $k \in \{h, v, d\}$ (detail coefficients) and $i \in \{1, \ldots, L\}$ (resolution level). The coarse approximation is represented as $\mathbf{c}_{\mathbf{o}a,L}[m, n]$.

In the second step, the watermark is embedded by quantizing the coefficients. For each bit that should be embedded, a key K determines the resolution level and the pixel position chosen for embedding. The detail coefficients at the selected resolution and position are sorted in ascending order: $\mathbf{c}_{\mathbf{o}k_1,l}[m, n] \leq \mathbf{c}_{\mathbf{o}k_2,l}[m, n] \leq \mathbf{c}_{\mathbf{o}k_3,l}[m, n]$. To embed the watermark, the range of values between $\mathbf{c}_{\mathbf{o}k_1,l}[m, n]$ and $\mathbf{c}_{\mathbf{o}k_3,l}[m, n]$ is divided into bins and $\mathbf{c}_{\mathbf{o}k_2,l}[m, n]$ is quantized, which is shown in Figure 4.4.

Finally, the inverse wavelet decomposition calculates the watermarked image data. For retrieval of the watermark, similar steps have to be performed. The detected watermark is evaluated with the embedded watermark by correlation. A similar algorithm was proposed by Ohnishi and Matsui [27].

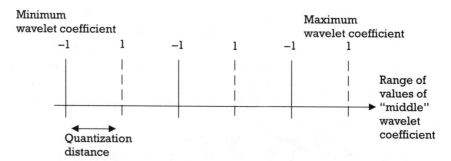

Figure 4.4 The quantization proposed by [26]. The three detail wavelet coefficients of a certain level and at a certain position are sorted. The information is embedded in the coefficient of the middle order by quantization.

Fractal watermarking schemes are based on fractal compression, which is developed from iterated function systems (IFS) and is based on self-similarity. More details on fractal image compression can be found in [28, 29]. The fractal encoding algorithm can be described as follows [30]: The original image is partitioned into nonoverlapping range cells. The image is covered with a sequence of possibly overlapping domain cells. For each range cell, the corresponding domain cell and the transform are searched to determine which covers the range cell best. This last step is computationally expensive. The range of transforms typically includes affine transforms and change of brightness and contrast. The transform found describes the self-similarity between a range cell and the corresponding domain cell. For embedding the watermark, the range cells are restricted by the information that should be encoded [31]. For retrieving the watermark from a marked range block, the corresponding domain cell reveals the embedded information. This is exemplified in Figure 4.5.

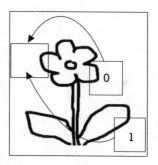

Figure 4.5 Example of embedding information by changing the self-similarity. The position of the similar block determines the value embedded.

Modification of geometric features is a method for watermarking images that is described by Rongen et al. [32]. They propose using a pseudorandomly generated dense line pattern which represents the watermark. Salient image points are detected, (e.g., as described by Shi and Tomasi [33]). The image is warped to move these salient points in the vicinity of the lines. During the detection of the watermark, the number of salient points in the vicinity is evaluated.

4.2.2 Watermarking methods dealing with geometric distortions

The general problem with the previously described watermarking methods is the proper synchronization needed for retrieving the watermark. Applying geometrical transforms to or warping the watermarked images affects this synchronization. Some watermarking algorithms, for example, block dependent algorithms, require a proper alignment, and therefore are not inherently robust against translation. Among these general transforms are global affine transforms. Yet local affine transforms and projective transforms also have to be considered. These geometrical transforms are considered desynchronization attacks (see Section 7.2.1) and are applied in image watermarking benchmarks (see Section 7.4.2), like the historic StirMark benchmark [34], for the evaluation of robustness. Furthermore, a combination of simple transforms influences many classical watermarking schemes drastically, although visible artifacts are hardly perceptible for images. Besides the exhaustive search of the embedded watermark, different strategies have been developed to address the problem of geometrical desynchronization attacks. In general, additional information is embedded. This additional information can be the redundant watermark content or additional information to recover the original geometry of the watermarked image.

Redundant embedding of the watermark content is proposed in different publications. For example, the watermark can be embedded periodically as proposed by Honsinger [35] or Kutter [36]. This periodic embedding results in the characteristic peaks in the autocorrelation function (ACF). These peaks reflect the applied geometrical transforms. However, an attacker can also calculate the ACF with the aim of predicting the watermark. Further publications based on redundant embedding use cyclic properties of the watermark pattern [37] or use redundant embedding in video sequences [38]. A watermarking method based on the autocorrelation function, which especially addresses local nonlinear geometrical distortions, was proposed by Voloshynovskiy et al. in [39, 40].

Invariant transform domains are applicable in increasing robustness. Some transforms are inherently not affected by specific geometrical transforms. For example, replacing the DCT transform with an invariant transform like log-polar mapping (LPM), which is also called the *Fourier-Mellin transform*, as described by Ó Ruanaidh and Pun [41], has some theoretical advantages. After applying the DFT, which is invariant to translation, every amplitude in the DFT at position (u, v) is projected in a new coordinate space (ρ, θ) via the projection:

$$u = e^{\rho} \cos(\theta) \tag{4.4a}$$

$$v = e^{\rho} \cos(\theta) \tag{4.4b}$$

In this new coordinate system, rotation and scaling are converted into translation. By calculating the amplitude of the DFT of the LPM, the resulting domain is invariant against rotation, scaling, and translation (RST). The embedding scheme is shown in Figure 4.6.

Because of practical problems, the authors suggested embedding the watermark into the translation-invariant DFT domain and adding a (second) template watermark into the RST-invariant LPM domain. Another approach that uses the properties of the LPM domain was proposed by Lin et al. [42, 43].

Template insertion is another technique for increasing the robustness of watermarking techniques. In the case of image watermarking, a template is inserted in the image. This template is used to recover the original image format and does not carry any watermarking content. One of those methods was proposed by Pereira and Pun [44]. The template consists of points that are randomly arranged in the DFT domain. Their radii vary between two limiting frequencies and are chosen (magnitude and phase) via a secret key. Peaks are generated by increasing the coefficients at the calculated positions. The watermark detection process consists of two steps. First, the template

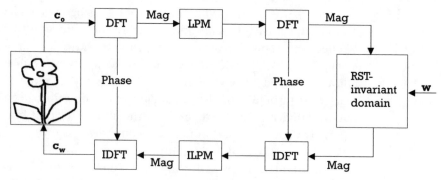

Figure 4.6 The LPM scheme for RST invariance as proposed by Ó Ruanaidh and Pun [41].

is detected. This information is used to calculate a linear transform. Second, the information about the linear transform is used to retrieve the embedded watermark. As with redundant embedding, an attacker can also use information about the template to attack the embedded watermark, as described by Herrigel et al. [45].

Further approaches have considered a number of properties for embedding, for example, geometry recovery by using the original watermarked image as proposed by [46, 47]. These methods require the original image instead of using a template. The original image is used to identify the geometrical distortions and to undo them. The main disadvantage is the fact that blind or oblivious detection is not possible with these methods after a geometrical attack. Using regions of interest (ROI) for watermarking as proposed by Su et al. [48] is currently difficult to achieve without human interaction because semantically meaningful regions have to be identified. However, content-based watermarking based on robust segmentation, as presented in the next section, is a generalized variant of watermarking of ROIs.

4.2.3 Content-based watermarking methods

While newer methods also have to face the previously described problem of geometrical distortions, they attempt to use semantic information in the image—the content of the image—for synchronization. Thus, they are classified as content-based watermarking algorithms.

In "Towards Second Generation Watermarking Schemes," Kutter et al. [49] outlined a scheme that is based on significant features concerning perception. These features should be invariant to noise, covariant to geometrical transforms, and independent of cropping. For feature extraction, the image is decomposed using the Mexican-Hat wavelet as proposed by Manjunath et al. [50]. The detected features are used for an image segmentation using Voronoi diagrams (i.e., partitioning of a given space). The resulting segments are used for embedding a watermark with an existing watermarking scheme. The detected feature is used as a reference origin for the watermarking process. For the detection of the watermark, the same features have to be extracted and the image has to be segmented. The authors reported that the feature location may move by 1 or 2 pixels, which has to be compensated for by a limited search. A similar scheme which is well described is presented by Bas et al. [51].

Instead of creating a triangulation of the image data, Dittmann et al. [52] proposed a scheme based on self-spanning patterns (SSP). These SSPs are also based on image feature points. The initial pattern, which is represented by a polygon, is spanned over four feature points. Information carrying patterns are spanned around the previous pattern, resulting in a set of polygons with a given traverse direction.

An estimation of images parameters is proposed by Alghoniemy and Tewfik [53] which is also based on the wavelet decomposition. Previous proposed methods like [54] suggested using image moments which are invariant against geometrical transforms [55]. However, their main disadvantage is the missing robustness against cropping, which is addressed by the method presented in [53]. The scaling parameter is estimated using the edges standard deviation ratio (ESDR) and the rotation angle is estimated using the average edges angles difference (AEAD). These estimations are based on the wavelet maxima which are extracted from the low-frequency components of the wavelet decomposition. The ESDR and the AEAD show increased robustness against cropping. However, they are not completely robust against general affine transforms. Therefore, Alghoniemy and Tewfik propose to combine this method with exhaustive search strategies.

Local watermarks are proposed by Tang and Hang [56]. Their scheme uses the same feature extraction method as proposed by Kutter et al. [49]. These extracted features build the centers of nonoverlapping image disks. The watermarks are embedded and extracted in each image disk. Before embedding and detection, the image disks are normalized using the normalization method proposed by Alghoniemy and Tewfik [53]. The watermark is then embedded in the DFT domain.

Segmentation or region-based image watermarking algorithms are proposed by Nikolaidis and Pitas [57] and by Celik et al. [58]. In contrast to the methods in which a region of interest has to be selected manually, these methods use image segmentation methods to group the pixels of an image according some statistics. The method proposed by Celik et al. [58] is based on a color clustering using a k-means clustering method. The cluster centers are identified and a Delauny triangulation of these cluster centers results in image regions that are watermarked. Thus, this watermarking technique is related to the triangulation method used in [51]. However, different features are used for the triangulation.

The method proposed by Nikolaidis and Pitas [57] is based on the iterated conditional modes (ICM) for clustering. The resulting regions are merged and the largest regions of the final results are used for embedding the watermark. Before watermarking, the regions are approximated by ellipsoids.

The bounding rectangle of each ellipsoid is used for the embedding and detection of the watermark.

4.3 Binary and halftoned images

Halftoning converts continuous-tone or multitone images into two-tone (binary) images. The visual result is supposed to be almost the same as the original gray-scale image. This is quite commonly used for printing books or newspapers. Watermarking methods for gray-scale images are generally not inherently robust against this type of conversion. Therefore, different watermarking techniques have been developed considering robustness against halftoning. In contrast to gray-scale images, binary images consist only of black and white pixels. Depending on the media type, different types of binary images can be distinguished depending on the degree of freedom of the pixels' perimeter and location.

While dot-area modulation (DAM) is often used to encode the visible image, dot-position modulation (DPM) can be used for hiding additional information as proposed by Rosen and Javidi [59]. Changing the area of dots is independent from changing the position, while the DPM-encoded image cannot be directly seen by an observer. For revealing the embedded information, a two-dimensional spatial correlator is proposed.

While the previous method was developed for printed images, the following method was developed for binary digital images: Changing the ratio of pixels was proposed by Zhao and Koch [60]. This method changes the ratio (percentage) of black and white pixels in a selected block. For each block, the percentage of black pixels is calculated: $P_{\text{black}}(\text{block}) = \frac{\#_{\text{black}}(\text{block})}{\#_{\text{pixels}}(\text{block})}$. The percentage of white pixels is given by $P_{\text{white}}(\text{block}) = 100\% - P_{\text{black}}(\text{block})$. Embedding of information is done by changing the ratio of the black or white pixels into a certain interval. This is a quantization of the ratio of black and white pixels. This quantization is shown in Figure 4.7. For modifying the thresholds, two different types of image areas are distinguished by Zhao and Koch: areas with well-distributed pixels and areas containing boundaries. For increased quality, different strategies for modifications are

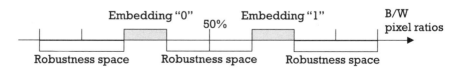

Figure 4.7 Embedding scheme for half-toned images using different dither matrices.

proposed: While pixel modifications are spread all over the area in dithered areas, pixel modifications are limited to the boundary in areas with boundaries. A similar method was presented by Wu et al. [61].

A kind of masking technique is suggested by Deseilligny and Le-Men [62]. The pixels that are flipped in the watermarked image depend on the original image and the embedded watermark. The embedded watermark is a binary image, for example, the logo of a company. This binary image is repeated periodically to cover the whole image that should be watermarked. White pixels in the original image flip if and only if the pixel of the watermark at the same position is black and a neighboring pixel in the original image (e.g., the left one) is black too. For recovering the size of the watermark, content (logo) must be known and the reverse strategy can be used for recovering the embedded information.

A simple method, which is called *data hiding in block parity* (DHBP), is proposed by Fu and Au [63], which is similar to the watermarking method of binary images proposed by Wu et al. [61]. The block-sum parity (even or odd number of pixels) encodes the information. The method proposed by Baharav and Shaked [64] uses different dither matrices (instead of one). The watermark determines the dither matrices used. The dither matrix influences the pixel distribution in the output image. The scheme is shown in Figure 4.8. For decoding the distance of the watermark, the dithered image (region) is used to determine the embedded value.

Binary text documents are considered by Mei et al. [65] (see Section 6.4.2). The interesting aspect of this approach is that different patterns in text documents (similar to strokes) are identified and modified. The modifications change the number of pixels, but the main properties remain the same. Thus, a maximum quality is achieved.

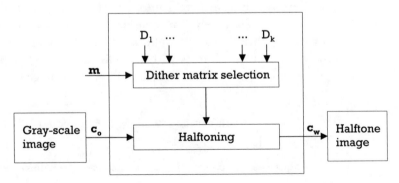

Figure 4.8 Quantization of the black and white pixel ratios as proposed by Zhao and Koch [60].

4.4 Summary

While some of the improved image watermarking techniques based on communication theory, especially those methods that are based on autocorrelation, show increased robustness against typical image processing operation combinations, they still have to prove their robustness against targeted attacks, as shown in Chapter 7. Embedding information by using image features for synchronization is promising, but these methods still have their individual weaknesses.

References

[1] Perry, B., B. MacIntosh, and D. Cushman, "Digimarc MediaBridge—The Birth of a Consumer Product, from Concept to Commercial Applicaton," in E. J. Delp and P. W. Wong, (eds.), *Proceedings of Electronic Imaging 2002, Security and Watermarking of Multimedia Contents IV*, San Jose, CA, January 2002, pp. 118–123.

[2] Fridrich, J., and M. Goljan, "Practical Steganalysis of Digital Images: State of the Art," in E. J. Delp and P. W. Wong, (eds.), *Proceedings of Electronic Imaging 2002, Security and Watermarking of Multimedia Contents IV*, San Jose, CA, January 2002, pp. 1–13.

[3] Voloshynovskiy, S. V., et al., "StegoWall: Blind Statistical Detection of Hidden Data," in P. W. Wong and E. J. Delp, (eds.), *Proceedings of Electronic Imaging 2002, Security and Watermarking of Multimedia Contents IV*, San Jose, CA, January 2002, pp. 57–68.

[4] van Schyndel, R. G., A. Z. Tirkel, and C. F. Osborne, "A Digital Watermark," *International Conference on Image Processing*, Austin, 1994, pp. 86–90.

[5] Westfeld, A., "F5-A Steganographic Algorithm," in I. S. Moskowitz, (eds.), *Information Hiding: 4th International Workshop*, Vol. 2137 of *Lecture Notes in Computer Science*, April 2001, Pittsburgh: Springer-Verlag, pp. 289–302.

[6] Lee, Y., and L. Chen, "A High Capacity Image Steganographic Model," *IEE Proceedings Vision, Image and Signal Processing*, Vol. 147, No. 3, June 2000, pp. 288–294.

[7] Crandall, R., "Some Notes on Steganography," 1998, http://os.inf.tu-dresden.de./~westfield/crandall.pdf.

[8] Cox, I. J., et al., "Secure Spread Spectrum Watermarking for Multimedia," *IEEE Transactions on Image Processing*, Vol. 6, No. 12, December 1997, pp. 1673–1687.

[9] Cox, I. J., et al., *Secure Spread Spectrum Watermarking for Multimedia*, Technical Report 95-10, NEC Research Institute, 1995.

[10] Cox, I. J., M. L. Miller, and J. A. Bloom, *Digital Watermarking*, The Morgan Kaufmann Series in Multimedia Information and Systems, San Francisco: Morgan Kaufmann Publishers, 2002.

[11] Hartung, F., and M. Kutter, "Multimedia Watermarking Techniques," *Proceedings of IEEE*, Vol. 87, No. 7, July 1999, pp. 1079–1107. (Special issue on Identification and Protection of Multimedia Information.)

[12] Langelaar, G. C., I. Setyawan, and R. L. Lagendijk, "Watermarking Digital Image and Video Data: A State-of-the-Art Overview," *Signal Processing Magazine*, Vol. 17, No. 5, September 2000, pp. 20–46. (Special issue on digital watermarking.)

[13] Jähne, B., *Digital Image Processing*, 3rd ed., Heidelberg: Springer-Verlag, 1995.

[14] Ó Ruanaidh, J. J. K., W. J. Dowling, and F. M. Boland, "Phase Watermarking of Digital Images," *Proceedings of the 1996 International Conference on Image Processing*, Vol. 3, Lausanne, September 1996, pp. 239–242.

[15] Barni, M., et al., "A DWT-Based Technique for Spatio-Frequency Masking of Digital Signatures," in P. W. Wong and E. J. Delp, (eds.), *Proceedings of Electronic Imaging '99, Security and Watermarking of Multimedia Contents*, Vol. 3657, San Jose, CA, January 1999, pp. 31–39.

[16] Stollnitz, E. J., T. D. DeRose, and D. H. Salesin, *Wavelets for Computer Graphics: Theory and Applications*, Morgann Kaufmann, 1996.

[17] Inoue, H., et al., "A Digital Watermark Technique Based on the Wavelet Transform and Its Robustness on Image Compression and Transformation," *Proceedings of the 1998 IEEE International Conference on Image Processing (ICIP-98)*, Vol. 2, Chicago, October 1998, pp. 391–395.

[18] Wang, H.-J. M., P.-C. Su, and C.-C. J. Kuo, "Wavelet-Based Digital Image Watermarking," *Optics Express 491*, Vol. 3, No. 12, December 1998.

[19] Meerwald, P., and A. Uhl, "A Survey of Wavelet-Domain Watermarking Algorithms," in P. W. Wong and E. J. Delp, (eds.), *Proceedings of Electronic Imaging 2001, Security and Watermarking of Multimedia Contents III*, San Jose, CA, January 2001, pp. 505–516.

[20] Pitas, I., and T. H. Kaskalis, "Applying Signatures on Digital Images," *IEEE WorkShop on Nonlinear Image and Signal Processing*, Neos Marmaras, Greece, June 1995, pp. 460–463.

[21] Langelaar, G. C., J. C. A. van der Lubbe, and J. Biemond, "Copy Protection for Multimedia Data Based on Labeling Techniques," in G. H. L. M. Heideman, (eds.), *17th Symposium on Information Theory in the Benelux*, Enschede, the Netherlands, May 1996, pp. 33–40.

[22] Bruyndonckx, O., J.-J. Quisquater, and B. Macq, "Spatial Method for Copyright Labeling of Digital Images," in I. Pitas, (eds.), *Nonlinear Signal Processing Workshop*, Halkidiki, Greece, June 1995, pp. 456–459.

[23] Burgett, S., E. Koch, and J. Zhao, "Copyright Labeling of Digitized Image Data," *IEEE Communications Magazine*, Vol. 36, No. 3, March 1998, pp. 94–100.

[24] Chen, B., and G. W. Wornell, "Quantization Index Modulation: A Class of Provably Good Methods for Digital Watermarking and Information

Embedding," *IEEE Transactions on Information Theory*, Vol. 47, No. 4, May 2001.

[25] Eggers, J. J., and B. Girod, "Blind Watermarking Applied to Image Authentication," *IEEE International Conference on Acoustics, Speech and Signal Processing (ICASSP)*, Salt Lake City, May 2001.

[26] Kundur, D., and D. Hatzinakos, "Digital Watermarking Using Multiresolution Wavelet Decomposition," *International Conference on Acoustics, Speech and Signal Processing (ICASSP)*, Seattle, May 1998, pp. 2969–2972.

[27] Ohnishi, J., and K. Matsui, "Embedding a Seal into a Picture under Orthogonal Wavelet Transform," *Proceedings of the 1996 International Conference on Multimedia and Computing Systems*, June 1996, pp. 514–522.

[28] Barnsley, M. F., and L. Hurd, *Fractal Image Compression*, Boston: Academic Press, 1992.

[29] Fisher, Y., (ed.), *Fractal Image Compression—Theory and Application*, New York: Springer-Verlag, 1995.

[30] Welstead, S., *Fractal and Wavelet Image Compression Techniques*, Bellingwood, WA: SPIE Optical Engineering Press, 1999.

[31] Puate, J., and F. Jordan, "Using Fractal Compression Scheme to Embed a Digital Signature into an Image," *Proceedings of SPIE Photonics East '96 Symposium*, Boston, November 1996.

[32] Rongen, P. M. J., M. J. B. Maes, and C. W. A. M. van Overveld, "Digital Image Watermarking by Salient Point Modification—Pratical Results," in P. W. Wong and E. J. Delp, (eds.), *Proceedings of Electronic Imaging '99, Security and Watermarking of Multimedia Contents*, San Jose, CA, January 1999, pp. 273–282.

[33] Shi, J., and C. Tomasi, *Good Features to Track*, Technical Report, TR93-1399, Cornell University Computer Science Department, November 1993.

[34] Petitcolas, F. A. P., R. J. Anderson, and M. G. Kuhn, "Attacks on Copyright Marking Systems," In D. Aucsmith, (eds.), *Information Hiding: Second International Workshop*, Vol. 1525 of *Lecture Notes in Computer Science*, Portland, OR: Springer-Verlag, April 1998, pp. 218–238.

[35] Honsinger, C., "Data Embedding Using Phase Disperson," *Proceedings of the Conference on Image Processing, Image Quality, Image Capture Systems (PICS-00)*, Springfield, VA, March 2000, pp. 264–268.

[36] Kutter, M., Watermarking Resisting to Translation, Rotation, and Scaling, "*Proceedings of SPIE: Multimedia Systems and Applications*, Boston, November 1998, pp. 423–431.

[37] Delanay, D., and B. Macq, "Generalized 2-D Cyclic Patterns for Secret Watermark Generation," *International Conference on Image Processing (ICIP-99)*, Vol. 2, September 2000, pp. 77–80.

[38] Kalker, T., et al., "A Video Watermarking System for Broadcast Monitoring," in P. W. Wong and E. J. Delp, (ed.), *Proceedings of Electronic Imaging '99, Security and Watermarking of Multimedia Content*, Vol. 3657, January 1999, pp. 103–112.

[39] Voloshynovskiy, S., F. Deguillaume, and T. Pun, "Multibit Digital Watermarking Robust Against Local Nonlinear Geometrical Distortions," *Proceedings of the IEEE International Conference on Image Processing (ICIP'01)*, Thessaloniki, Greece, October 2001, pp. 1023–1026.

[40] Deguillaume, F., S. Voloshynovskiy, and T. Pun, "Method for the Estimation and Recovering from General Affine Transforms in Digital Watermarking Applications," in P. W. Wong and E. J. Delp, (eds.), *Proceedings of Electronic Imaging 2002, Security and Watermarking of Multimedia Contents IV*, San Jose, CA, January 2002, pp. 491–502.

[41] Ó Ruanaidh, J. J. K., and T. Pun, "Rotation, Scale and Translation Invariant Digital Image Watermarking," *Proceedings of the 1997 International Conference on Image Processing*, Vol. 1, Santa Barbara, CA, October 1997, pp. 536–539.

[42] Lin, C.-Y., et al., "Rotation, Scale, and Translation Resilient Watermarking for Images," *IEEE Transactions on Image Processing*, Vol. 10, No. 5, May 2001.

[43] Lin, C.-Y., et al., "Rotation, Scale, and Translation Resilient Public Watermarking for Images," in P. W. Wong and E. J. Delp, (eds.), *Proceedings of Electronic Imaging 2000, Security and Watermarking of Multimedia Contents II*, San Jose, CA, 2000, pp. 90–98.

[44] Pereira, S., and T. Pun, "Fast Robust Template Matching for Affine Resistant Image Watermarking," in A. Pfitzmann, (ed.), *Information Hiding, 3rd International Workshop*, Vol. 1768 of *Lecture Notes in Computer Science*, Dresden: Springer-Verlag, September 1999, pp. 340–353.

[45] Herrigel, A., S. Voloshynovskiy, and Y. Rytsar, "The Watermark Template Attack," in P. W. Wong and E. J. Delp, (eds.), *Proceedings of Electronic Imaging 2001, Security and Watermarking of Multimedia Contents*, San Jose, CA, January 2001, pp. 394–405.

[46] Johnson, N. F., Z. Duric, and S. Jajodia, "Recovery of Watermarks from Distorted Images," in A. Pfitzmann, (ed.), *Information Hiding, 3rd International Workshop*, Vol. 1768 of *Lecture Notes in Computer Science*, Dresden: Springer-Verlag, October 1999, pp. 318–332.

[47] Sun, Q., J. Wu, and R. Deng, "Recovering Modified Watermarked Image with Reference to Original Image," in P. W. Wong and E. J. Delp, (eds.), *Proceedings of Electronic Imaging '99, Security and Watermarking of Multimedia Contents*, San Jose, CA, January 1999, pp. 415–424.

[48] Su, P.-C., H.-J. Wang, and C.-C. J. Kuo, "Digital Watermarking in Regions of Interest," *IS&T Image Processing/Image Quality/Image Capture Systems (PICS)*, Savannah, GA, April 1999.

[49]   Kutter, M., S. K. Bhattacharjee, and T. Ebrahimi, "Towards Second Generation Watermarking Schemes," *Proceedings of the 1999 International Conference on Image Processing (ICIP-99)*, Kobe, Japan, October 1999, pp. 320–323.

[50]   Manjunath, B., C. Shekhar, and R. Chellappa, "A New Approach to Image Feature Detection with Applications," *Pattern Recognition*, Vol. 29, 1996, pp. 627–640.

[51]   Bas, P., J.-M. Chassery, and B. Macq, "Geometrically Invariant Watermarking Using Feature Points," *IEEE Transactions on Image Processing*, Vol. 11, No. 9, September 2002, pp. 1014–1028.

[52]   Dittmann, J., T. Fiebig, and R. Steinmetz, "New Approach for Transformation-Invariant Image and Video Watermarking in the Spatial Domain: Self-Spanning Patterns (SSP)," *Proceedings of Electronic Imaging 2000, Security and Watermarking of Multimedia Contents II*, San Jose, CA, January 2000, pp. 176–185.

[53]   Alghoniemy, M., and A. H. Tewfik, "Geometric Distortion Correction in Image Watermarking," *Proceedings of Electronic Imaging 2000, Security and Watermarking of Multimedia Contents II*, Vol. 3971, San Jose, CA, January 2000.

[54]   Alghoniemy, M., and A. H. Tewfik, "Geometric Distortion Correction Through Image Normalization," *Proceedings of the IEEE International Conference on Multimedia and Expo (ICME)*, Vol. 3, New York, August 2000, pp. 1291–1294.

[55]   Shen, D. G., and H. S. H. Ip, "Generalized Affine Invariant Image Normalization for Rotational Symmetric and Non-Rotational Symmetric Shapes," *IEEE Transactions on Pattern Analysis and Machine Intelligence (PAMI)*, Vol. 19, No. 5, May 1997, pp. 431–440.

[56]   Tang, C.-W., and H.-M. Hang, "Image-Feature-Based Robust Digital Watermarking Scheme," in E. J. Delp and P. W. Wong, (eds.), *Proceedings of Electronic Imaging 2002, Security and Watermarking of Mulitmedia Contents IV*, San Jose, CA, January 2002, pp. 584–595.

[57]   Nikolaidis, A., and I. Pitas, "Region-Based Image Watermarking," *IEEE Transactions on Image Processing*, Vol. 10, No. 11, November 2001, pp. 1726–1740.

[58]   Celik, M. U., et al., "Analysis of Feature-Based Geometry-Invariant Watermarking," in P. W. Wong and E. J. Delp, (eds.), *Proceedings of Electronic Imaging 2001, Security and Watermarking of Multimedia Contents*, Vol. 4314, San Jose, CA, June 2001, pp. 261–268.

[59]   Rosen, J., and B. Javidi, "Hidden Images in Halftone Pictures," *Applied Optics*, Vol. 40, No. 20, July 2001, pp. 3346–3353.

[60]   Zhao, J., and E. Koch, "Embedding Robust Labels into Images for Copyright Protection," in K. Brunnstein and P. P. Sint, eds., *Intellectual Property Rights and New Technologies. Proceedings of the KnowRight '95 Conference*, Vienna, Austria, 1995, pp. 242–251.

[61] Wu, M., E. Tang, and B. Liu, "Data Hiding in Digital Binary Images," *Proceedings of the 2000 IEEE International Conference on Multimedia and Expo (ICME 2000)*, New York, July 2000, pp. 393–396.

[62] Pierrot-Deseilligny, M., and H. Le-Men, "An Algorithm for Digital Watermarking of Binary Images: Application to Map and Text Images," *International Workshop on Computer Vision*, Hong Kong, September 1998.

[63] Fu, M. S., and O. C. Au, "Data Hiding Watermarking for Halftone Images," *IEEE Transactions on Image Processing*, Vol. 11, No. 4, 2002, pp. 477–484.

[64] Baharav, Z., and D. Shaked, "Watermarking of Dither Halftoned Images," in P. W. Wong and E. J. Delp, (eds.), *Proceedings of Electronic Imaging '99, Security and Watermarking of Multimedia Contents*, Vol. 3657, San Jose, CA, January 1999, pp. 307–316.

[65] Mei, Q., E. K. Wong, and N. Memon, "Data Hiding in Binary Text Documents," in E. J. Delp and P. W. Wong, (eds.), *Proceedings of Electronic Imaging 2001, Security and Watermarking of Multimedia Contents III*, San Jose, CA, January 2001, pp. 369–375.

Digital watermarking for audio data

Although the number of publications in the field of digital watermarking has increased approximately exponentially from an initial number of two in 1992, the majority of the published literature deals with the marking of image and video data, whereas publications in the audio-watermarking domain represent a relatively low percentage that is steadily increasing since 2000. This chapter presents several techniques and approaches for a realization of watermarking systems in the audio domain. From the variety of existing algorithms, a few categories of methods can be identified according to certain aspects built into the different schemes. Besides the general requirements valid for the watermarking of every media type, audio data–related requirements and corresponding design criteria will be discussed in Section 5.1. Preserving the quality of the watermarked audio tracks requires the integration of psychoacoustic models, which are presented in Section 5.2, followed by approaches to integrate them in a watermarking encoder in Section 5.3. Section 5.4 presents several algorithms that form the main classes of research in the audio watermarking field.

5.1 Requirements and design

According to the intended application of watermarks in audio data, the algorithm as well as the watermark itself has to fulfill a set of requirements [1]. The IFPI has specified the desired features of an optimal audio watermarking method. These requirements can be elaborated and subdivided further into signal processing properties, security properties, and application-specific requirements of the algorithm and the watermark.

Quality and robustness are related to the properties of the watermarked tracks and the embedded watermarks, and are general requirements for all watermarking systems. Nevertheless the ranking of these two requirements is special in the audio case (see below). The catalogue of possible audio signal manipulations depending on the application contains but is not limited to the following signal manipulations [2], which can be grouped into different categories (see Section 7.2.1).

5.1.1 Removal manipulations and attacks

- Addition of multiplicative and additive noise;
- Filtering like low-, high-, and allpass filtering;
- Lossy compression, for example, MPEG audio layer I, II, III;
- Noise reduction applying different kinds of algorithms;
- Digital to analog (D/A) and analog-to-digital (A/D) conversion;
- Changing the sampling rate (i.e., quantization of the audio track);
- Collusion and statistical attacks.

5.1.2 Misalignment manipulations and attacks

- Fluctuating time and pitch scaling;
- Cropping or insertion of samples.

Besides removal and misalignement, embedding and detection attacks can be applied as in all other watermarking domains (see Section 7.2.1). In addition to the general requirements of the quality of the watermarked copies and the robustness and security of the embedded watermarks, applications like active broadcast monitoring and customer identification for transaction tracking extend the range of the necessary features of the underlying algorithm. Both types of applications have in common the fact that the watermark is a function of time only known right before the time of delivery. Therefore, the speed of the watermark encoder is of vital importance for the applicability of the watermarking techniques. Even for real-time watermarking systems, the need to embed a large number of different watermarks simultaneously is a critical issue.

The requirements detailed above describe the maximum sets of criteria an audio watermarking alogrithm has to fulfill. Since the described features can in general not be fulfilled simultaneously in each imaginable application, different variations and corresponding design criteria are relevant for the development of an effective method. The most important requirement addresses the quality of the watermarked audio tracks. If the quality of the audio tracks cannot be preserved, neither users (whether consumers or broadcast industry professionals) nor especially the recording industry will accept this technology. This emphasizes the priority in ranking among the requirements from quality (first) to robustness (second) and data capacity (third).

To ensure the quality of the watermarked audio tracks, a psychoacoustic model has to be an integral part of the watermark encoder. The next section will deal with the psychoacoustic phenomena and models used in current high-quality audio watermarking systems.

5.2 Psychoacoustic facts and models

The science of psychoacoustics describes acoustics from the perspective of the human auditory system. The abilities of the auditory system are not only investigated as the qualitative relation between sound and the corresponding impression, but also as quantitative relations between the stimuli presented and hearing sensations [3]. Digital audio technologies like the development of MPEG rely on the detailed knowledge of the human auditory system.

The relevant information is not only limited by the ability of the ear to hear frequencies in a band between 20 Hz and 20 kHz and the dynamic range of over 96 dB, but the interaction of different frequencies and the corresponding processing of the human auditory system is also important to consider for a deeper understanding of the correlation between acoustical stimuli and hearing sensations.

The development of an exact model of the auditory system is a complex and, to a certain extent, subjective task. Physically, sounds are easily described by the time-varying sound pressure $p(t)$. The processing of this sound pressure leads to a complex auditory sensation. The input to the human auditory system are the temporal variations in sound pressure. The processing in the auditory system leads to an output that contains information about the temporal and spectral characteristics of sound as well as the localization of the sound source.

The science of psychoacoustics tries to describe this information processing of the human auditory system. The most significant recent contributions in this field were made by Zwicker and Fastl [3]. As indicated above, the development of sophisticated models for the auditory system is based in large part on the use of extensive experimental data which of necessity imply a certain subjective aspect that can only be compensated through a sufficiently large data set.

Masking, pitch, critical bands, excitation, and just-noticeable changes describe the active processing of the ear. From these effects, masking plays the most important role in the lossy compression of digital audio data. Sophisticated models of masking in the frequency and time domain have been developed and applied to effective compression of audio data [4]. Data reduction by a factor of 12 can be achieved without a significant loss of sound quality.

5.2.1 Critical bands

Critical bands are an important concept in describing the auditory sensations. A corresponding construct, a so-called critcal-band rate scale, was defined which is based on the fact that the human auditory system analyzes a broad spectrum into different parts. These parts are the so-called critical bands. Table 5.1 was built by adding one critical band to the next in such a way that the upper limit of the lower critical band corresponds to the lower limit of the next higher critical band.

The critical bandwidth has a constant value of 100 Hz up to a center frequency of approximately 500 Hz. Above 500 Hz, a good approximation for the bandwidth Δf_G/Hz is 20% of the actual frequency. The following two analytic expressions are used to describe the dependence of critical band

Table 5.1 Critical-Band Rate z and Corresponding Frequencies

z/bark	f_u/Hz	f_o/Hz	Δf_G/Hz	z/bark	f_u/Hz	f_o/Hz	Δf_G/Hz
0	0	100	100	13	2,000	2,320	320
1	100	200	100	14	2,320	2,700	380
2	200	300	100	15	2,700	3,150	450
3	300	400	100	16	3,150	3,700	550
4	400	510	110	17	3,700	4,400	700
5	510	630	120	18	4,400	5,300	900
6	630	770	140	19	5,300	6,400	1,100
7	770	920	150	20	6,400	7,700	1,300
8	920	1,080	160	21	7,700	9,500	1,800
9	1,080	1,270	190	22	9,500	12,000	2,500
10	1,270	1,480	210	23	12,000	15,500	3,500
11	1,480	1,720	240	24	15,500		
12	1,720	2,000	280				

rate and critical bandwidth over the entire auditory frequency range:

$$z = 13 \ \arctan\left(0.76\tfrac{f}{\text{kHz}}\right) + 3.5 \ \arctan\left(\tfrac{f}{7.5 \ \text{kHz}}\right)^2 \ [\text{Bark}] \qquad (5.1)$$

$$\Delta f_G = 25 + 75\left[1 + 1.4\left(\tfrac{f}{\text{kHz}}\right)^2\right]^{0.69} \ [\text{Hz}] \qquad (5.2)$$

The critical bandwidth as a function of frequency shows the nonlinear behavior of our hearing system. The critical band rate grows from 0 to 24 and has the unit Bark. The critical band rate is related to several other scales that describe characteristics of the hearing system.

5.2.2 Masking effects

Masking is an effect that occurs in everyday life. To enable a normal conversation, the power of speech does not need to be very high. However, if emergency vehicles are passing by with loud sirens while we are talking on the street, our conversation is nearly impossible. We normally have to wait until the emergency vehicle has passed or raise our voice to a greater loudness in order to continue our conversation. These effects also take place in the case of music, where louder instruments can mask out faint ones. This is a typical example of the so-called simultaneous masking. Nonsimultaneous masking takes place when the masker and the test sound are not presented simultaneously but in close connection in time.[1] Two different situations are distinguished according to time relation of the test sound and the masker:

1. Premasking[2] occurs when the test sound is presented before the masker.

2. Postmasking, also called *backward masking*, takes place when the test sound is presented after the masker is no longer present.

Besides this total masking of sound, there exists also a so-called partial masking effect which reduces the loudness of the test sound. Since this effect is not relevant in the case of audio watermarking, we will not consider it more closely. In order to measure the effect of masking quantitatively, the so-called masking threshold is determined. It is the sound pressure level of the test tone necessary to be just audible if a masker is applied. It is identical

---

1. Masker is the tone which masks out other sound. Test sound is the sound that will be masked.
2. The expressions prestimulus and forward masking are also used.

with the threshold in quiet if the frequencies of the masker and the test tone are very different.

The simultaneous and nonsimultaneous masking phenomena are segregated by the temporal characteristics of the masker and test sound. Steady-state conditions can be assumed if the test and masking sound have a duration longer than 200 ms.

Summary of masking effects   Nearly every type of music has a strong temporal structure. Test and masking sound having a temporal characteristic produce so-called temporal masking effects. In order to measure the time relations between test tone and masker, the test sound is shifted relative to the masker. According to the time shift Δt relative to the masker, three different regions can be differentiated (see Figure 5.1).

The premasking effect happens before the masker is switched on in region I. Premasking lasts about 20 ms in any condition. This means the threshold remains unchanged until Δt reaches a negative value of 20 ms according to Figure 5.1. After $-20 \, \text{ms} \leq \Delta t$, the threshold increases and reaches the level found in simultaneous masking just before the masker is switched on. The effect of premasking looks like listening into the future at a first glance. Obviously, the information processing in our auditory system does not work instantaneously. The time needed to perceive the sound depends on the

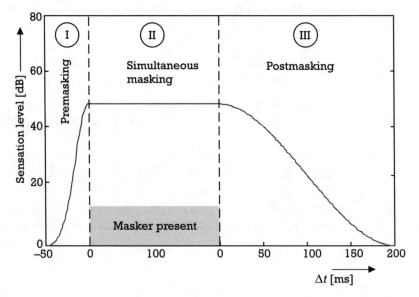

Figure 5.1   Regions with premasking, simultaneous masking, and postmasking.

loudness of the presented sound. Therefore, loud maskers have a shorter setup time than the faint test sound and will be perceived earlier.

The second region is the area of simultaneous masking. Threshold in quiet and masked thresholds depend on the duration of the test sound. This can be explained if one takes into account that the hearing system integrates the sound intensity over a period of 200 ms. Therefore, for durations of the test sound shorter than 200 ms, the threshold in quiet and the masking thresholds increase because of the lower intensity due to the integration ability of the auditory system. The third region describing postmasking corresponds to a decay of the masking effect after the masker is switched off. After a 5-ms delay the masking threshold decreases from the level it had in the simultaneous masking region. At about 200 ms, the level of the masking threshold reaches the threshold in quiet. Postmasking depends strongly on the duration of the masker. The decay of the masking threshold is much steeper for shorter maskers.

5.2.3 Psychoacoustic model MPEG 1 layers I and II

Psychoacoustic models used in current audio compression encoders apply the frequency and temporal masking effects in order to ensure inaudibility by shaping the quantization noise according to the masking threshold. In turn, a natural approach is to use already existing models for shaping the watermark noise. The different psychoacoustic models differ in complexity and the implementation of the different masking effects. One of the frequently used models is the psychoacoustic model 1 layers I and II of ISO-MPEG with $f_s = 44.1$ kHz [4–6]. It supports the sampling rates $f_s = 32, 44.1$ and 48 kHz. In order to iteratively allocate the necessary bits, the MPEG standard calculates the signal-to-mask ratios (SMR) of all the subbands. This requires the determination of the maximum signal level and the minimum masking threshold in each subband.

Calculation of the power density spectrum   To derive the masking threshold, the power density spectrum of the input block has to be estimated. In order to minimize the leakage effect, the input block is multiplied with a Hanning window defined by

$$h(i) = \sqrt{\frac{8}{3}}\left[1 - \cos\left(\frac{2\pi i}{N}\right)\right], \quad i = 0, 1, \ldots, N-1 \qquad (5.3)$$

Layer I uses an input block $s(l), l = 1 \ldots N$ of length $N = 512$, whereas Layer II is operating on blocks with $N = 1{,}024$ samples.

After multiplication with the Hanning window, the FFT of the input block is performed.

$$X(k) = 10 \log_{10} \left| \frac{1}{N} \sum_{l=0}^{N-1} h(l)s(l)e^{-i\frac{2\pi kl}{N}} \right|^2 \text{ [dB]} \qquad (5.4)$$

The maximum value of the power density spectrum X is normalized to a value of $+96$ dB.

Determination of sound pressure level   The sound pressure level L_s in band n is calculated by

$$L_s(n) = \max\{X(k), 20 * \log_{10}[\text{scf}(n) * 2^{15}] - 10\} \text{ [dB]}$$

$$\text{with } X(k) \text{ in subband } n \qquad (5.5)$$

$X(k)$ is the result from the power density spectrum calculation. $\text{scf}(n)$ is the scaling factor in subband n. After the segmentation of the frequency bands, this factor is determined from the maximum value of 12 successive samples in band n via a lookup table. According to that lookup table, the scaling factor only determines the peak level for a period of time. The multiplication by a factor of 2^{15} is the normalization to $+96$ dB. The -10-dB term corrects the difference between peak and root-mean-square (RMS) level (see [4]).

Threshold in quiet   The threshold in quiet LT_q (also called *absolute threshold*) is defined as the sound pressure level of a pure tone that is just audible as a function of the frequency [3]. The following expression can be used as an approximation:

$$LT_q = 3.64 \left(\frac{f}{\text{kHz}} \right)^{-0.8} - 6.5e^{-0.6(\frac{f}{\text{kHz}} - 3.3)^2} + 10^{-3} \left(\frac{f}{\text{kHz}} \right)^4 \text{ [dB]} \qquad (5.6)$$

The normalization of the absolute threshold in quiet is done by adjusting the function according to the following rule: A signal with a frequency of 4 kHz and an amplitude[3] of ± 1 LSB lies on the curve of the absolute threshold [7]. The absolute threshold is available in the form of tables for the different sampling rates f_s (see [4]). In order to take the threshold in quiet into account in the calculation of the global masking threshold, the tables contain values for all frequencies necessary to compute the masking threshold. An additional

---

3. Amplitude resolution is 16 bits.

corrective offset has to be added depending on the overall bit rate used per channel.

$$\text{Offset} = \begin{cases} -12\text{dB}, & \text{bit rate} < 96 \text{ Kbps} \\ 0\text{dB}, & \text{bit rate} \geq 96 \text{ Kbps} \end{cases} \tag{5.7}$$

If the computed masking threshold lies below the threshold in quiet, the masking threshold is set to the absolute threshold in each band.

Determination of tonal and nontonal components   The masking curves are determined by the tonality of the individual masker. Therefore, the discrimination between the different components has to be performed. The first step is a determination of the local maxima in the power density spectrum (see Figure 5.2).

$$X(k-1) < X(k) \quad \text{and} \quad X(k) \geq X(k+1) \tag{5.8}$$

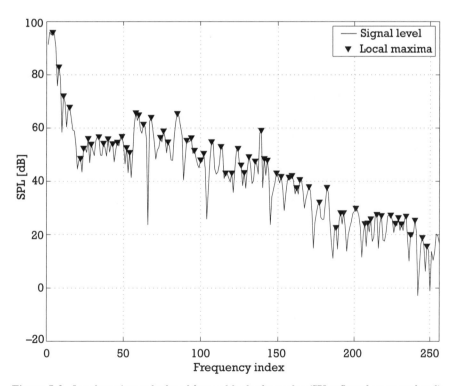

Figure 5.2   Local maxima calculated from a block of samples (SPL = Sound pressure level).

After this step, the determination of the tonal components within the bandwidth of a critical band is based on the examination of the power of neighboring spectral lines. A local maximum $X(k)$ is also a tonal component if the following criterion is met[4]:

$$X(k) - X(k - j) \geq 7[\text{dB}] \tag{5.9}$$

with

$$
\begin{array}{lll}
j = & -2, +2 & \text{for} \quad 2 < k < 63 \\
j = & -3, -2, +2, +3 & \text{for} \quad 63 \leq k < 127 \\
j = -6, \ldots, -2, +2, \ldots, +6 & \text{for} \quad 127 \leq k < 250
\end{array} \tag{5.10}
$$

The sound pressure level of the tonal maskers are computed as given in (5.11):

$$X_{tm}(k) = 10 \log_{10} \left(10^{\frac{X(k-1)}{10}} + 10^{\frac{X(k)}{10}} + 10^{\frac{X(k+1)}{10}} \right) [\text{dB}] \tag{5.11}$$

The nontonal components are computed from the remaining lines without the tonal components within each critical band. The power of these spectral lines is summed to form the nontonal component $X_{nm}(k)$ corresponding to the critical band.[5] The index k of the nontonal component is defined by the index of the spectral line nearest to geometric mean of the critical band. Figure 5.3 displays a power density spectrum with the tonal and nontonal components found for a block of $N = 512$ samples and sampling rate $f_s = 44.1$ kHz.

Decimation of tonal and nontonal components   The number of maskers considered for the calculation of the global masking threshold is reduced in this step. Components are removed from the list of relevant components if their power values are below the absolute threshold.

$$X_{tm}(k) \geq LT_q(k) \quad \text{or} \quad X_{nm}(k) \geq LT_q(k) \tag{5.12}$$

For the tonal components, an additional decimation is performed if two or more components are separated by less than 0.5 Bark. The tonal component with the highest power is kept, whereas all other components are removed

---

4. k is the frequency index.
5. The indexes nm and tm denote the nontonal and tonal maskers, respectively.

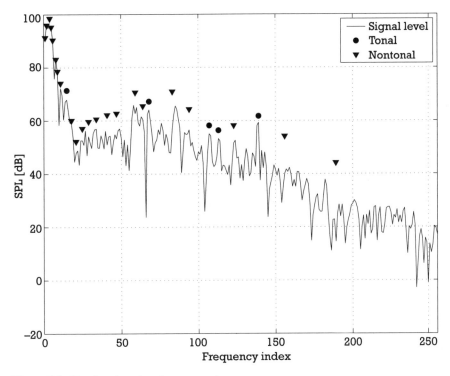

Figure 5.3   Tonal and nontonal components.

from the list of tonal components. This operation is performed by applying a sliding window of width 0.5 Bark in the critical band. The remaining tonal and nontonal components are used in the calculation of the individual masking thresholds.

Calculation of individual masking thresholds LT_{tm} and LT_{nm}   The MPEG model uses only a subset of the $N/2$ spectral lines to calculate the global masking threshold. The reduction to the subsampled frequency domain is a nonlinear mapping of the $N/2$ frequency lines. The number of samples used in the subsampled frequency domain varies depending on the sampling rate and layers. For Layer I, samples numbers are

$$f_s = 32\,\text{kHz} \qquad n = 108 \tag{5.13a}$$

$$f_s = 44.1\,\text{kHz} \qquad n = 106 \tag{5.13b}$$

$$f_s = 48\,\text{kHz} \qquad n = 102 \tag{5.13c}$$

The masking thresholds for tonal and nontonal maskers can be calculated using (5.14):

$$LT_{tm}[z(j), z(i)] = X_{tm}[z(j)] + av_{tm}[z(j)] + vf[z(j), z(i)][dB] \quad (5.14)$$

$$LT_{nm}[z(j), z(i)] = X_{nm}[z(j)] + av_{nm}[z(j)] + vf[z(j), z(i)] [dB] \quad (5.15)$$

The masking threshold is calculated at the frequency index i, while j is the frequency index of the masker and $X_{tm}[z(j)]$ is the power density of the masker with index j. The term av is the so-called *masking index* and vf the *masking function*. Here $z(j)$ is the so-called *critical band rate* and denotes the frequency in the Bark scale. The bark values and the corresponding frequency indexes are stored in tables. Figure 5.4 displays the decimated tonal masker and the corresponding individual masking thresholds.

The masking index for tonal and nontonal masker can be calculated by

$$av_{tm}[z(j)] = -1.525 - 0.275 * z(j) - 4.5[dB] \quad (5.16)$$

$$av_{nm}[z(j)] = -1.525 - 0.175 * z(j) - 0.5[dB] \quad (5.17)$$

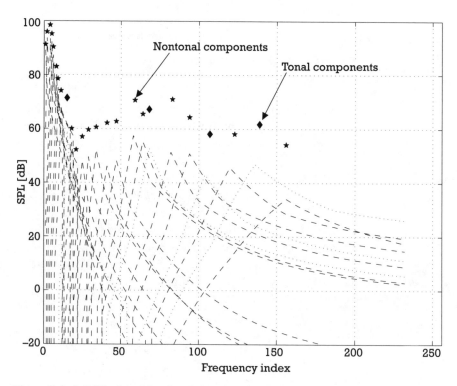

Figure 5.4 Individual masking thresholds for the tonal components.

The masking function vf$[z(j), z(i)]$ with the distance in Bark $\Delta z = z(i) - z(j)$ is defined by (5.18):

$$
\text{vf} = \begin{cases}
17(\Delta z + 1) - (0.4X[z(j)] + 6) & -3 \leq \Delta z < -1 \\
(0.4X[z(j)] + 6) * \Delta z & -1 \leq \Delta z < 0 \\
-17\Delta z & 0 \leq \Delta z < 1 \\
-(\Delta z - 1) * (17 - 0.15X[z(j)]) - 17 & 1 \leq \Delta z < 8 \\
\text{in [decibels]} & \text{in Bark}
\end{cases}
\qquad (5.18)
$$

Calculation of the global masking threshold LT_g   In order to calculate the global masking threshold LT_g, the different components have to be summed up. The global masking thresholds for frequency index i are computed by adding the power of the threshold in quiet and the tonal and nontonal masker in each case (see Figure 5.5).

$$
LT_g(i) = 10 \log_{10}\left(10^{\frac{LT_q(i)}{10}} + \sum_{j=1}^{N_t} 10^{\frac{LT_{tm}[z(i),z(j)]}{10}} + \sum_{j=1}^{N_n} 10^{\frac{LT_{nm}[z(i),z(j)]}{10}} \right) [\text{dB}]^6 \quad (5.19)
$$

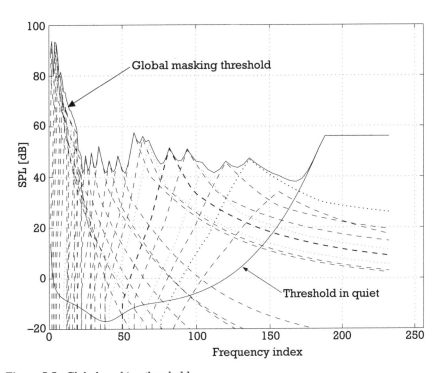

Figure 5.5   Global masking threshold.

6. N_t and N_n denote the number of tonal and nontonal components.

Computation of the minimum masking threshold LT_{Min}   The global masking threshold LT_g is computed in the subsampled frequency domain with the number of spectral lines according to (5.13). These frequency indexes are mapped onto the 32 subbands.

$$LT_{\mathrm{Min}}(n) = \min_{f(i)\in\text{ subband } n} LT_g(i)\,[\mathrm{dB}] \tag{5.20}$$

Calculation of the signal-to-mask ratio   The SMR is calculated for every subband n:

$$\mathrm{SMR}(n) = L(n) - LT_{\mathrm{Min}}(n)\,[\mathrm{dB}] \tag{5.21}$$

The output of the psychoacoustic model is the SMR. This information is used by lossy audio compressors to iteratively allocate the bits in every subband. This is not necessary in the case of a watermarking application, since only the masking threshold for each block is of interest. Therefore, the integration of the psychoacoustic model requires only the following steps:

1.   Calculation of the power spectrum;

2.   Identification of the tonal (sinusoid-like) and nontonal (noise-like) components;

3.   Decimation of the maskers to eliminate all irrelevant maskers;

4.   Computation of the individual masking thresholds;

5.   Computation of the global masking threshold;

6.   Determination of the minimum masking threshold in each subband.

5.3   Perceptual audio watermarking

The task of a watermark encoder is to adjust the watermark signal in order to ensure inaudibility and simultaneously embed the watermark with the maximum power according to the carrier signal to provide maximum robustness. A perceptual audio watermarking encoder typically consists of several components (see Figure 5.6). The encoding and modulation block encode the information **m** by means of a secret key K and modifies selected carrier components of the audio signal like the amplitude, phase, and frequency magnitude according to the underlying algorithm. The psychoacoustic model (PAM) block analyzes the original signal $c_o(t)$ in order to calculate perception thresholds like the minimum masking threshold LT_{Min}. It can also represent the psychoacoustic control parameters like the maximum allowable

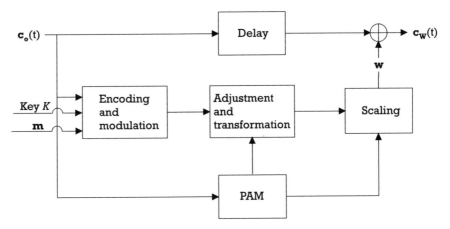

Figure 5.6 Watermark encoder and its components.

phase difference (see Section 5.4.2) or temporal masking thresholds (see Figure 5.1). The model used is driven by the kind of modulation used in the specific algorithm.

Since the majority of perceptual audio watermarking algorithms are embedding the watermark in the Fourier domain, we will consider the spectral shaping of the watermark noise more closely.

5.3.1 Spectral weighting

In this case, the adjustment block in Figure 5.6 represents a spectral weighting block multiplying the frequency components of the watermark noise with the weight coefficients calculated by the PAM block. The spectral weighting is often implemented as a filter block on a piecewise linear approximation of the masking threshold which represents the frequency response. Therefore, the original has to be delayed before being added to the watermark noise. The filter used in approximating LT_{Min} can be a finite impulse response (FIR) filter designed with the window method. The signal loss due to the window edges can be minimized by overlapping successive blocks by 50% (see [1]). These calculations have to be performed for each block of length $\Delta t = \frac{512 \text{ samples}}{f_s \text{ samples}/s}$ with sampling frequency f_s.

Scaling of watermark noise level   After filtering the watermark noise, the sound pressure level has to be adjusted. To calculate the correct attenuation of the watermark noise, the power spectrum is estimated. The masking threshold is the result of the psychoacoustic analysis. It is based on a normalized original signal with a maximum shifted to +96 dB. This normalization has to be taken into account in the scaling block if one calculates the power of the watermark noise relative to the original signal (see Figure 5.7).

The so-called *noise-masking-ratio* NMR $= W - LT_{\text{Min}}$ [dB] serves as an additional attenuation factor which adjusts the level of quality against robustness. The result is the actual watermark signal added to the original signal to produce the watermarked track $c_{\mathbf{w}}(t)$.

After presenting the underlying psychoacoustic principles and methods to integrate them into audio watermarking algorithms, the next section presents a variety of algorithms developed in recent years.

5.4 Algorithms

A variety of approaches already exist to embed information into audio data. The techniques range from the simple LSB method to the spread-spectrum methods. Features used for embedding the watermark bits are the amplitude in the time domain; magnitude, frequency and phase in the Fourier domain; and characteristics of the compressed audio stream.

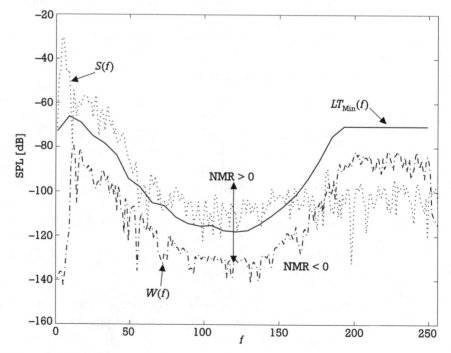

Figure 5.7 Sound pressure levels.

In the notation used throughout this chapter, $c_o[i], i = 1, \ldots, l(c_o)$ are the samples of the original signal in the time domain.[7] The range of sequence of numbers is according to the amplitude resolution of 8- or 16-bit $c_o[i] \in \{0, 255\}$ or $c_o[i] \in \{-32,768, +32,767\} \forall i$. An additional index of the carrier elements c_{oj} denotes a subset of the audio signal. To the authors' best knowledge, all audio watermarking algorithms split the audio signal into different overlapping or nonoverlapping blocks.[8] Therefore, $c_{oj}[i]$ denotes the ith sample in the jth block of size $l(c_{oj})$. The individual blocks are used to embed part of 1 bit of information, 1 bit, a sequence of bits, or the whole watermark denoted by m. The length of the blocks is often determined by the usage of psychoacoustic models (see Section 5.2.3), the special transformation performed on the block, or the number of bits to be embedded.

5.4.1 LSB coding

One of the first techniques investigated in the watermarking field, as for virtually all media types, is the so-called LSB encoding (see Section 2.3.3). It is based on the substitution of the LSB of the carrier signal with the bit pattern from the watermark noise. It uses no psychoacoustic model in order to shape the watermark. A natural approach in the case of audio data is to alter the LSB of the individual samples of the digitized audio stream having an amplitude resolution of, for example, 16 bits.[9] This *blind watermarking* method requires an exact synchronization of the marked audio data during the detection procedure. Besides having a high payload of 44.1 Kbps, its low robustness makes these algorithms useless in real watermarking applications.

5.4.2 Embedding watermarks into the phase

Approaches that embed the watermark into the phase of the original signal do not use the temporal or spectral masking effects (see Section 5.2), but exploit the fact that the human auditory system has a low sensibility against relative phase changes [3].

5.4.2.1 Phase coding

The method presented by Bender et al. [8] splits the original audio stream into blocks and embeds the whole watermark into the phase spectrum of the first block. The original signal c_o is split into $M = \left\lfloor \frac{l(c_o)}{N} \right\rfloor$ blocks $c_{oj}, 0 \leq j \leq M-1$ with $N := 2l(m)$ samples.

---

7. If the audio data is sampled at a sampling rate $f_s = 44.1$ kHz, 1 sec corresponds to $l(c_o) = 44,100$.

8. Usually with the same size.

9. The usual amplitude resolution for audio files in CD format.

1. Each block of $\mathbf{c_o}$ is transformed in the Fourier domain $\mathbf{C_o}_j = \mathcal{F}\{\mathbf{c_o}_j\}$, $\forall j$. A matrix of the phases $\phi_{oj}[\omega_k]$ and magnitudes $|\mathbf{A_o}_j[\omega_k]|$, $0 \le k \le N/2 - 1$ is constructed.

2. The matrix with the differences in phase between the M neighbor blocks is computed:

$$\Delta\phi_{oj+1}[\omega_k] = \phi_{oj+1}[\omega_k] - \phi_{oj}[\omega_k], \forall j, k \qquad (5.22)$$

3. The watermark is encoded in the phase spectrum of the first block:

$$\phi_{w0}[\omega_k] = (-1)^{\mathbf{m}[k]+1}\frac{\pi}{2}, \quad \text{for} \quad \mathbf{m}[k] \in \{0, 1\}, \ 0 \le k \le N/2 - 1 \qquad (5.23)$$

4. In order to ensure the inaudibility of the phase changes between the individual blocks, the phase differences in each of the blocks have to be adjusted:

$$\phi_{wj+1}[\omega_k] = \phi_{wj}[\omega_k] - \Delta\phi_{oj+1}[\omega_k], \forall j, k \qquad (5.24)$$

5. The original magnitudes $|\mathbf{A_o}|$ and the modified phase spectrum ϕ_w of the blocks are used to compute the marked signal in the time domain $\mathbf{c_w}_j = \mathcal{F}^{-1}\{\mathbf{C_w}_j\}\forall j$.

Before decoding the watermark, a preprocessing step is necessary in order to synchronize with the beginning of the starting sequence. A necessary precondition in the decoding stage is the knowledge of the length[10] of the watermark $l(\mathbf{m})$.

1. Synchronization onto the first block $\mathbf{c_{w0}}$;

2. Transformation of the block $\mathbf{C_{w0}} = \mathcal{F}\{\mathbf{c_{w0}}\}$;

3. Reading the bits of the watermark from the phase information of the first block $\phi_{w0}[\omega_k]$ for $0 \le k \le N - 1$.

One disadvantage of the phase coding approach is the low payload that can be achieved. Only the first block is used in embedding the watermark.

---

10. The watermark length defines the number of samples via $N = 2l(\mathbf{m})$ used in the DFT.

Moreover, the watermark is not distributed over the entire data set, but is implicitly localized and can thus be removed easily if cropping is acceptable.

5.4.2.2   Phase modulation

Another form of embedding the watermark into the phase is by performing independent multiband phase modulation [9]. Inaudible phase modifications are exploited in this algorithm by controlled multiband phase alterations of the original signal. The original signal $\mathbf{c_o}$ is segmented into $0 \leq m \leq M - 1$, $M = \lceil \frac{l(\mathbf{c_o}) - N}{2N} \rceil$, blocks with N samples using overlapping windows. The window function is

$$\text{win}[n] = \sin\left(\frac{\pi(2n+1)}{2N}\right), \quad 0 \leq n \leq N - 1 \tag{5.25}$$

Two adjacent blocks consist of the original and a watermarked block. The kth watermarked block ($k = 2m$) carries the kth sequence of the watermark. To ensure inaudibility by introducing only small changes in the envelope, the phase modulation is performed by fulfilling the following constraint given in (5.26):

$$\left|\frac{\Delta\phi(z)}{\Delta z}\right| < 30° \tag{5.26}$$

where $\phi[z]$ denotes the signal phase and z is the Bark scale according to (5.1). A slow phase change over time is achieved by using a long block size of $N = 2^{14}$.

1.   Each block of $\mathbf{c_o}$ to be watermarked is transformed in the Fourier domain $\mathbf{C}_{ok} = \mathcal{F}\{\mathbf{c}_{ok}\}$, $k = 2m$, $1 \leq m \leq \lfloor \frac{M-1}{2} \rfloor$ yielding the Fourier coefficients $\mathbf{A}_{ok}[f]$.

2.   The next step constructs the phase modulation function $\Phi_k(b)$. One integer Bark scale carries 1 message bit of the watermark. Each message bit is represented by a phase window function centered at the end of the corresponding Bark band and spans 2 Bark bands.

$$\phi(z) = \sin^2\left(\frac{\pi(z+1)}{2}\right), \quad -1.0 \leq z < 1.0 \tag{5.27}$$

The sign $\mathbf{a}_k[j] \in \{-1, 1\}$ of the phase window function is determined by the jth message bit $\mathbf{m}_k[j] \in \{0, 1\}$ of the kth sequence. The total phase modulation is obtained by the linear combination of the overlapped phase window functions:

$$\Phi_k(z) = \sum_{j=1}^{J} \mathbf{a}_k[j]\phi(z-j), \quad 0.0 \le z < J \tag{5.28}$$

3. Using the $\Phi_k(z)$, the bits are embedded into the phases in the kth audio block by multiplying the Fourier coefficients with the phase modulation function:

$$\mathbf{A}_{\mathbf{w}k}[f] = \mathbf{A}_{\mathbf{o}k}[f] \times e^{i\Phi_k[f]} \tag{5.29}$$

with f the frequency in hertz in contrast to the Bark scale z [see (5.1)].

4. The marked signal is computed by inverse transformation of the modified Fourier coefficients $\mathbf{A}_{\mathbf{w}k}$ of the individual blocks $\mathbf{c}_{\mathbf{w}k} = \mathcal{F}^{-1}\{\mathbf{C}_{\mathbf{w}k}\}, \forall k$. All blocks are windowed and overlap-added to create the watermarked signal.

The robustness of the modulated phase can be increased by using n_z Bark values carrying 1 message bit. Since one integer Bark carries 1 message bit, the increase can be calculated by $\pm 15 \times n_z$ corresponding to (5.26). For audio tracks sampled with sampling rate f_s and the number of critial bands N_B, the data rate is[11]

$$\#\text{bits per block} \times \#\text{blocks per second} = \frac{N_B}{n_z} \times \frac{f_s}{N} \tag{5.30}$$

Retrieving the watermark requires a synchronization procedure to perform a block alignment for every watermarked block by using the original signal. The watermark bits from the kth audio block are recovered from the obtained phase modulation $\widehat{\Phi}_k$ for that block. A matching of the individual segments of the modulated phase to the encoded bits would be possible if the phase modulation is not distorted by manipulations. In contrast, the retrieved $\widehat{\Phi}_k$ is a noisy version of the modulated phase Φ_k, preventing an easy decoding of the kth sequence of bits from the kth audio block. Nevertheless, according to (5.28), the modulated phase Φ_k can be viewed as a sequence of state transitions of the four possible transitions $(0 \to 0, 0 \to 1, 1 \to 0, 1 \to 1)$. Besides decoding each bit individually,

---

11. $N_B = 24$ for $f_s = 44.1$ kHz according to Table 5.1.

this enables the modelling of $\widehat{\Phi}_k$ as a hidden Markov model and a determination of the single best concatened sequence of those possible transitions. The possible transitions $\mathbf{p}^{ij}[f] = \{p_t^{ij}[f]\}_{t=1}^T, 0 \leq i, j \leq 1$ can be calculated in advance for the frequency range (in hertz) used for embedding the individual bit. T is determined by the number of bits embedded and f covers the frequency range in hertz for the tth bit.

1. Calculate phase modulation function $\widehat{\Phi}_k$ by applying the window function (5.25) and performing the Fourier transformation $\mathbf{C}_{\mathbf{w}k} = \mathcal{F}\{\mathbf{c}_{\mathbf{w}k}\}$ of the kth block.

2. Formulate the $\widehat{\Phi}_k$ as an observation sequence $\mathbf{o}[f] = \{o_t[f]\}_{t=1}^T$, where f covers the frequency range in hertz for the tth bit.

3. Calculate the weight factor sequence $\beta[f] = \{\beta_t[f]\}_{t=1}^T$ with

$$\beta_t[f] = \min\left(\left|\mathbf{A}_{\mathbf{w}t}[f]\right|^2, \left|\mathbf{A}_{\mathbf{o}t}[f]\right|^2\right), \quad \text{for } f = 0, \dots, K-1 \quad (5.31)$$

$$\sum_f \beta_t[f] = 1 \qquad (5.32)$$

The weight factors of the tth observation sequence are determined by the smaller spectrum energy of the original or watermarked signal. This is based on the assumption that smaller spectrum components and their corresponding phase information are more likely to be distorted by some kind of nonlinear processing like MPEG encoding.

4. Calculate the cost function

$$c_t^{ij} = \frac{1}{K} \sum_{f=0}^{K-1} \left|(p_t^{ij}[f] - o_t[f])\beta_t[f]\right|, \quad \text{for} \quad 0 \leq i, j \leq 1, 1 \leq t \leq T$$

$$(5.33)$$

5. Perform the Viterbi [10] search algorithm with the calculated cost function in order to find the best sequence of possible state transitions, which in turn yields the kth sequence of bits.

Both phase embedding approaches use the psychoacoustic features of the human auditory system with regard to the just noticeable phase changes. They exploit the inaudibility of phase changes if the time envelope of the original signal is approximately preserved. Because of the phase alteration, embedding and detection of the watermark is done in the Fourier domain by processing the audio stream blockwise. While the phase coding method is embedding the watermark in the phases of the first block, the phase modulation algorithm performs a long-term multiband phase modulation. Both algorithms are *nonblind watermarking* methods, since they require the original signal during the watermark retrieval, which of course limits their applicability.

5.4.3   Echo hiding

A variety of watermarking algorithms [11–15] are based on so-called *echo hiding* methods. Echo hiding algorithms embed watermarks into a signal $c_o(t)$ by adding echos $c_o(t - \Delta t)$ to produce a marked signal $c_w(t)$:

$$c_w(t) = c_o(t) + \alpha c_o(t - \Delta t) \tag{5.34}$$

Equation (5.34) contains two parameters that can be changed in order to provide inaudibility of the watermark and to embed the bits into the audio signal. The change of the delay time Δt is used to encode the bits of the watermark, whereas both parameters α and Δt have to be adjusted to ensure inaudibility of the embedded echo. In general, (5.34) can be written as

$$c_w(t) = \sum_{k=0}^{N} \alpha_k c_o(t - \Delta t_k) \tag{5.35}$$

where $c_o(t)$ is the original signal with $\alpha_0 = 1$, $\Delta t_0 = 0$, and N the number of different echo signals embedded. Using the response function

$$h(t) = \sum_{k=0}^{N} \alpha_k \delta(t - \Delta t_k) \tag{5.36}$$

this can be written in short form as a convolution of these echoes with the original signal

$$c_w(t) = c_o(t) * h(t) \tag{5.37}$$

In turn, the marked signal $c_w(t)$ can be expressed in the frequency domain as

$$C_w(\omega) = C_o(\omega) H(\omega) \tag{5.38}$$

where $C_o(\omega)$ and $H(\omega)$ are the Fourier transformations of the signals $c_o(t)$ and $h(t)$, respectively. During the detection step, the calculation of $h(t)$ is necessary in order to determine the individual echoes with corresponding delay times Δt_k encoding the bits $k = 1, \ldots, N$. According to (5.38), the signal can be deconvolved by dividing $C_w(\omega)$ by $C_o(\omega)$ in the frequency domain and calculating the inverse Fourier transformation. Performing this operation requires an a priori knowledge of the original signal $C_o(\omega)$, which is not practical in the case of watermarking. Therefore, the detection method uses the so-called *homomorphic deconvolution* technique in order to separate the signal and the echoes.

The basic idea behind homomorphic deconvolution is to apply a logarithmic function to convert the product (5.38) into a sum. Using the definition of the *complex cepstrum* as the inverse Fourier transformation of the log-normalized Fourier transform of the watermarked signal, the transformed signal can be written as

$$
\begin{aligned}
C_w(q) &= \mathcal{F}^{-1}\{\log|C_o(\omega)H(\omega)|\} \\
&= \mathcal{F}^{-1}\{\log|C_o(\omega)|\} + \mathcal{F}^{-1}\{\log|H(\omega)|\} \qquad (5.39) \\
&= C_o(q) + H(q)
\end{aligned}
$$

as a function of the time or *quefrency*[12] domain. According to (5.39), the original signal $C_o(q)$ and the embedded echos $H(q)$ are clearly separated on the quefrency axis q. Using this deconvolution technique in the detection of the watermark bits, an algorithm adding two different echoes for embedding 0 and 1 bits can be constructed. The original signal $\mathbf{c_o}$ is split into $M = \left\lfloor \frac{l(\mathbf{c_o})}{N} \right\rfloor$ blocks $\mathbf{c_o}_j$, $0 \le j \le M - 1$ with N samples. Each block carries 1 bit of the watermark.

1. For each block $\mathbf{c_o}_j$ of the original signal, the echo signal for the 0 and 1 bits are constructed with the corresponding delay time and attenuation factors α_0 and α_1.

$$
w_k(t) = \alpha_k c_o(t - \Delta t_k), \quad \text{for} \quad k = 0, 1 \qquad (5.40)
$$

2. Two complementary modulation signals $m_k(t)$, $k = 0, 1$ for the 0 and 1 bits are generated:

$$
m_0(t) = \sum_{j=0}^{M-1}(1 - b_j)\,\text{rect}_j(t), \quad m_1(t) = \sum_{j=0}^{M-1} b_j\,\text{rect}_j(t) \qquad (5.41)
$$

---

12. q is the quefrency and has the same units as time.

with

$$m_0(t) + m_1(t) = 1 \, \forall t \quad \text{rect}_j(t) = \begin{cases} 1 & \text{for} \quad t_j \leq t < t_{j+1} \\ 0 & \text{otherwise} \end{cases} \quad (5.42)$$

$$\text{and } b_j = \mathbf{m}[j \bmod l(\mathbf{m})]$$

The modulation signals are used to construct the echo signals according to the bits of the watermark.

3. After multiplying the echo signals $w_k(t)$ with the modulation signals $m_k(t)$, the marked audio stream is generated by addition of the computed signals to the original one:

$$c_w(t) = c_o(t) + m_0(t)w_0(t) + m_1(t)w_1(t) \quad (5.43)$$

Retrieving the watermark requires a synchronization procedure to perform an alignment with the watermarked blocks:

1. Transformation of the sequence in the cepstrum domain $\mathbf{C_w} = \mathcal{F}^{-1}\{\log(|\mathcal{F}\{\mathbf{c_w}\}|)\}$;

2. Autocorrelation of $\mathbf{C_w}$ in the cepstrum domain;

3. Measurement of the delay time δt via the peaks of the autocorrelation of $\mathbf{C_w}$;

4. Determination of the embedded bit by comparison of δt with Δt_k, $k = 0$ or 1.

From the masking effects presented in Section 5.2.2, the echo hiding approach uses the postmasking effect in order to control the inaudibility of the embedded watermark. The delay times Δt_k and attenuation factors α_k, $k = 0, 1$ have to be adjusted in the embedding process according to the perception threshold of the human auditory system (see Figure 5.1) to ensure the inaudibility of the echoes. It is a *blind watermarking* method, which modulates the bits as echo signals embedded in individual blocks of the audio stream. In contrast to the majority of audio watermarking algorithms, the embedding and the detection are performed in two different domains, the time and cepstrum domain, respectively. A disadvantage is the complexity of this method due to the number of transformations [see (5.39)] that have to be computed for detection, which is performed in the cepstrum domain. Furthermore one major drawback of this approach is the vulnerability to malicious attacks, since the information can be detected by anyone without using a secret key. An attacker can exploit this knowledge if he knows the underlying algorithm to apply a removal attack (see Section 7.2.2), which was demonstrated by Petitcolas et al. in [16]. A possible countermeasure presented by Ko et al. [13] against the easy determination of delay time δt in the detection procedure (see above) is the spreading of the echo over the time axis. This is accomplished by substituting the Dirac delta function in the response function (5.36) with a pseudonoise (PN) sequence. Instead of calculating the autocorrelation in the cepstrum domain, despreading of the echo is performed by cross-correlation of the cepstral signal (5.39) with the PN sequence generated from a secret key.

5.4.4 Watermarking of compressed audio data

5.4.4.1 Watermarking the compressed bit stream

Several approaches exist to embed the watermark directly into the already compressed audio bit stream (see [17–20]). The main argument for using such methods is that a lot of audio tracks, which are already published on the Internet, are compressed versions of the original one. Therefore, time-consuming decoding, watermarking embedding, and reencoding in the case of pulse code modulation (PCM) watermarking techniques are not necessary in order to embed the watermark. Furthermore, the retrieval process does not involve a decoding procedure, which results in an additional decrease in watermark retrieval speed. Nevertheless, the starting point for professionally created audio material is always the PCM format.[13] These approaches change the contents of the MPEG frame (see Figure 5.8) directly.

The scaling factor can be viewed as a logarithmic gain factor for the sample values in order to retrieve the original samples in PCM format. The embedding of the watermark is done by changing the scaling factors of different

Header 12-bit sync signal 20-bit system information	Error correction code 16-bit optional	Code for number of scale factors 2-bit	Bit assignment 4-,3-,2- bit for lower, middle, upper band	Scale factors 6-bit	Sample values 2-...15-bit	Optional data

Figure 5.8 MPEG-frame layer III.

---

13. Besides live recordings occasionally made on MiniDisc or sound tracks on MiniDV video camcorders.

frames according to a special pattern derived from a secret key. A problem of this method is that some audio streams carry only a few scaling factors per frame. Therefore, the space for embedding a watermark is reduced. This leads to the problem that multiple watermarks cannot be embedded, because altering scale factors already used for embedding the first watermark destroys the quality of the audio data.

A second approach in the variation of the MPEG frame [17] tries to alter the sample values instead of the scaling factors. Embedding multiple watermarks is also critical in this case. The additional requirement of using the original track as input for the retrieval process further limits the applicability of this approach.

Besides working on MP3 bit streams, methods like the one presented by Cheng et al. [20] are embedding watermarks into the advanced audio coding (AAC)[21, 22] compressed bit stream by direct modification of the quantized coefficients. The watermark bits are embedded by performing a spread-spectrum modulation (see Sections 2.3.3 and 5.4.5) of the quantized coefficients. The individual bits are retrieved by a linear correlation of the PN sequence used during the embedding and the quantized coefficients of the watermarked bit stream. The coefficients to be modified are selected by applying a heuristic, which uses only nonzero coefficients in a predefined frequency range. The amount of distortion applied is fixed and set to the quantization step size of 1.

Methods of directly watermarking the compressed bit stream have in common that they do not make use of a psychoacoustic model. Both embedding and detection are performed directly on the compressed bit stream, where the audio stream is processed in frames according to the formatting of the bit stream in the specific compression algorithm. Additional information is not necessary if the audio data are synchronized. The main advantage is the low computational cost. Furthermore, these methods obey implicit robustness against their specific compression format due to embedding of the watermark in the already compressed bit stream. The main disadvantage of these methods is the missing psychoacoustic counterpart in comparison to the uncompressed audio signal. The influence on the audio quality of the original track by altering scaling factors, sample data, or the quantized coefficients can only be estimated. Moreover, the decoding of the compressed bit stream and a new compression with a shifted audio stream may lead to a synchronization problem because of the new scaling factors, sample data, and quantization coefficients of the MPEG frames. Furthermore, the complexity advantage is lost if the watermarked audio tracks have to be transcoded in another compression format.

5.4.4.2 Integrating watermark embedding into compression encoder

Besides directly watermarking the bit stream, other methods extract the information in the compressed bit stream from the quantization of the audio samples [23]. This enables the estimation of the masking threshold to shape the watermark noise below this threshold in order to ensure inaudibility. Integrating the watermark and compression encoder has two advantages: The quality during the watermarking can be controlled in contrast to the methods described above and the speed of embedding is improved in comparison to two separate processes of watermarking and compression. The building blocks consists of parts of the PCM watermark embedder and the compression decoder and encoder (see Figure 5.9).

Part of the bit stream decoder is used in order to read the scaling factors and decode the bit stream and perform the inverse quantization of compressed samples. The information about the quantization enables the calculation of the masking threshold. The masking threshold controls the multiplication factors used to multiply the spectral lines of the constructed watermark— as usual in a perceptual watermark encoder—applying the masking effects. The watermark generation can be the same as for the PCM watermark embedder. After weighting the spectrum of the watermark noise, the result is added to the original spectral lines. The extracted scaling factors from the original frame are used in order to quantize the marked audio data again and format the bit stream. The final output is the marked bit stream (see Figure 5.10).

This method makes implicit usage of the psychoacoustic model by approximating the perceptual information contained in the MPEG frames. Detection can be performed on the compressed and uncompressed audio data. It is a *blind watermarking* method, which distributes the bits over different MPEG frames. Due to the usage of parts of the compression encoder and decoder, such a mechanism is tied to the special compression scheme used. For each newly developed compression algorithm, a new integration of the watermarking embedding procedure becomes necessary.

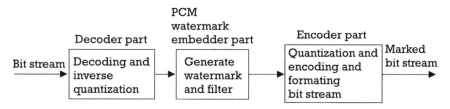

Figure 5.9 Integration of watermark embedding into compression encoder.

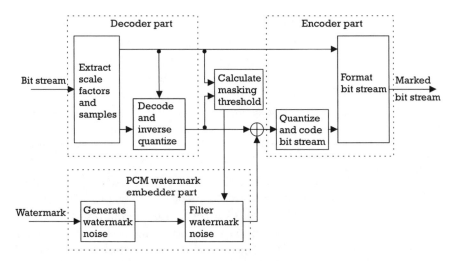

Figure 5.10   Components of bit stream watermarker.

5.4.5   Spread-spectrum audio watermarking

Spread-spectrum methods, originally conceived for masking the origin of radio transmissions and enhancing resilience against jamming, are often used in the transmission of digital information. Since the requirements of suppressing jamming during transmission, hiding a signal against an unintended listener, and ensuring information privacy are very similar to those in watermarking applications, these are probably the most widley used techniques in the development of watermarking algorithms. From the spread-spectrum viewpoint, the original audio signal can be considered as a jammer interfering with the signal carrying the watermark information (see Section 2.3.3).

The spread-spectrum modulation is a special form of watermark modulation. The modulation is performed on $\mathbf{C_o}$, which is the transformed block of samples $\mathbf{c_o}$. The transformation is used to model the audio signal with orthonormal base functions spanning the signal space. If the identity transformation is used, the signal is represented by the block of PCM samples itself. In the case of the Fourier transformation, the trigonometric functions are used as basis functions and the transformed block consists of the Fourier coefficients represented by the vector $\mathbf{C_o}$. Each bit $k \in \{0, 1\}$ is modeled by a pseudonoise $\mathbf{pn}_k$ vector consisting of two equiprobable elements $\{-1, +1\}$ generated by means of the secret key. Therefore, the expectation value of the pseudonoise sequence is $E\{\mathbf{pn}_k\} = 0$. Usually the pseudonoise sequences for the two bits are inverted $\mathbf{pn}_0 = -\mathbf{pn}_1 = \mathbf{pn}$. The original signal $\mathbf{c_o}$ is split into $M = \left\lfloor \frac{l(\mathbf{c_o})}{N} \right\rfloor$ blocks $\mathbf{c_o}_j$, $0 \leq j \leq M - 1$ with N samples.

To simplify the discussion, we consider one block ($\mathbf{c_o} := \mathbf{c}_{oj}$) carrying 1 bit of the watermark.[14]

1. The block $\mathbf{c_o}$ is transformed with the orthogonal transform T in the corresponding domain $\mathbf{C_o}$.

$$\mathbf{C_o} = T(\mathbf{c_o}) \tag{5.44}$$

2. The PN sequence $\mathbf{pn}_k$ is weighted with α to adjust between quality and robustness.

$$\mathbf{W} = \alpha\mathbf{pn}_k \tag{5.45}$$

3. The modulated and weighted watermark signal is added to the cover signal in the transformed domain.

$$\mathbf{C_w} = \mathbf{C_o} + \mathbf{W} \tag{5.46}$$

4. The watermarked signal is transformed back into the time domain.

$$\mathbf{c_w} = T^{-1}(\mathbf{C_w}) \tag{5.47}$$

During the detection step, the same vector $\mathbf{pn}_k, k = 0, 1$ has to be generated via the secret key. A comparator function is used in order to decide about the presence of the embedded vector $\mathbf{pn}$. This requires a perfect synchronization with the embedding block of samples.

1. Synchronization with the beginning of the embedding block $\mathbf{c_w}$;

2. Transformation of $\mathbf{c_w}$ into embedding domain $\mathbf{C_w} = T(\mathbf{c_w})$;

3. Correlation of $\mathbf{C_w}$ with $\mathbf{pn}_k, k = 0, 1$ by applying the comparator function C_τ (see Section 2.3.3):

$$C_\tau(\mathbf{C_w}, \mathbf{pn}) = C_\tau(\mathbf{C_o}, \mathbf{pn}) + C_\tau(\alpha\mathbf{pn}, \mathbf{pn}) \tag{5.48}$$

4. Detection of the transmitted bit, usually made on the sign of the comparator function

$$\text{sign}(C_\tau(\mathbf{C_w}, \mathbf{pn})) \begin{cases} > 0, & \text{for } \mathbf{pn}_0 \\ < 0, & \text{for } \mathbf{pn}_1 \end{cases} \tag{5.49}$$

14. The pattern can also be distributed over several blocks.

One of the widely used comparator functions C_τ is the linear correlation

$$C_\tau(\mathbf{x}, \mathbf{y}) = \langle \mathbf{x}, \mathbf{y} \rangle = \frac{1}{N} \sum_{i=1}^{N} \mathbf{x}[i]\mathbf{y}[i] \tag{5.50}$$

with the signal vectors $\mathbf{x}$ and $\mathbf{y}$. The result of the correlation consists of the two contributions $C_\tau(\mathbf{C_o}, \mathbf{pn})$ and $C_\tau(\alpha\mathbf{pn}, \mathbf{pn})$. The second term accumulates the contribution of the pseudonoise sequence[15] embedded in the different base functions, whereas the first term represents the correlation or the interference of the carrier signal respectively and pseudonoise sequence. If the pseudonoise sequence is split into the two sequences containing positive and negative elements, the correlation $C_\tau(\mathbf{C_o}, \mathbf{pn})$ can also be written as

$$C_\tau(\mathbf{C_o}, \mathbf{pn}) = \frac{1}{N} \sum_{i=1}^{N/2} \mathbf{C_o^+}[i] - \mathbf{C_o^-}[i] = \frac{(\mu^+ - \mu^-)}{2} \tag{5.51}$$

with μ^+ and μ^- denoting the mean values. According to the central limit theorem, the distribution of the means is normal if N is sufficiently large. Furthermore, the difference of two normal distributions is also normal with $N(\mu_{C_\tau}, \sigma_{C_\tau})$. Since $\mathbf{C_o}$ and $\mathbf{pn}$ are two independent random variables, the mean μ_{C_τ} and the variance σ_{C_τ} can be calculated according to

$$\mu_{C_\tau} = E\{C_\tau(\mathbf{C_o}, \mathbf{pn})\} = E\{\mathbf{C_o}\}E\{\mathbf{pn}_k\} = 0 \tag{5.52}$$

$$\sigma_{C_\tau}^2 \approx \hat{\sigma}_{\frac{(\mu^+ - \mu^-)}{2}}^2 = \hat{\sigma}_{\frac{(\mu^+ + \mu^-)}{2}}^2 = \hat{\sigma}_{\mu_{C_o}}^2 = \frac{\hat{\sigma}_{C_o}^2}{N} \tag{5.53}$$

By using the model of the distribution function $N(0, \sigma_{C_o}/\sqrt{N})$ in the unwatermarked case and assuming a fixed weighting $\alpha := \{\alpha\}_{i=1}^{N}$ of the pseudonoise sequence, the probability distribution function for the two different sequences is

$$f_{\mathbf{pn}_1}(t) = \frac{1}{\sqrt{2\pi}\sigma_{C_\tau}} e^{-\frac{(t-\alpha)^2}{2\sigma_{C_\tau}^2}} \quad f_{\mathbf{pn}_0}(t) = \frac{1}{\sqrt{2\pi}\sigma_{C_\tau}} e^{-\frac{(t+\alpha)^2}{2\sigma_{C_\tau}^2}} \tag{5.54}$$

---

15. This is often denoted as despreading the sequence.

Errors in detection of the bits occur if $C_\tau(\mathbf{C_o}, \mathbf{pn}) > C_\tau(\alpha \mathbf{pn}, \mathbf{pn})$. Therefore, the false alarm probability is obtained by

$$P_{fa} = P_{01} + P_{10} = p_0 \int_\tau^{+\infty} f_{\mathbf{pn}_0}(t)dt + p_1 \int_{-\infty}^\tau f_{\mathbf{pn}_1}(t)dt \qquad (5.55)$$

where P_{01} represents the error that a 0 bit is transmitted and a 1 bit is detected and P_{10} accordingly. Setting the a priori probabilites that the different bits are transmitted to $p_0 = p_1 = \frac{1}{2}$ and using the definition for the complementary error function $\mathrm{erfc}(x)$,

$$\mathrm{erfc}(x) = 1 - \mathrm{erf}(x) = \frac{2}{\sqrt{\pi}} \int_x^{+\infty} e^{-t^2} dt \qquad (5.56)$$

this can be written with the threshold $\tau = \alpha$ according to (5.55) as

$$P_{fa} = P_{01} + P_{10} = \frac{1}{2} \mathrm{erfc}\left(\sqrt{\frac{N}{2}} \frac{\alpha}{\sigma_{\mathbf{C_o}}} \right) \qquad (5.57)$$

Different kinds of audio watermarking algorithms use different embedding domains and representations of the transformed signal vector $\mathbf{C_o}$. Furthermore, the psychoacoustic parameters have to correspond to the specific embedding domain in order to perform the psychoacoustic weighting step. One of the first algorithms that used the masking properties human auditory system by Tewfik et al. [1, 2] works in the Fourier domain. The psychoacoustic weighting is performed by shaping the Fourier coefficients of the PN sequence according to the masking threshold calculated by the psychoacoustic model presented in Section 5.2.3. Furthermore, this algorithm approximates the temporal masking behavior (see Figure 5.1) by using the envelope of the signal for the increase and a decaying exponential for the decrease of the signal. Another algorithm presented by Haitsma et al. [24] also embeds the watermark in the Fourier domain by altering the frequency magnitudes. The algorithm presented by Kirovski and Malvar [25] uses the modulated complex lapped transform (MCLT)[16] and modifies the magnitude of the MCLT coefficients in the decibel scale rather than the linear scale. They use a psychoacoustic model [26], which quantifies the audibility of the MCLT magnitude coefficient. The algorithm proposed by Bassia and Pitas [27] works in the time domain by altering the amplitudes of the samples. The shaping of the watermark is applied by performing a lowpass filtering of the PN sequence.

16. The MCLT transformation is a 2x oversampled DFT filter bank.

Spread spectrum is a widely used technique for different types of media given its high robustness against signal manipulations. If a secret key is used to generate the pseudonoise sequence **pn**, this algorithm does not need the original audio signal in order to detect the embedded bits and is therefore a *blind watermarking* method, provided that the synchronization requirement is met. The main disadvantage is the vulnerability against desynchronization attacks, which will be addressed in Section 5.4.7. Furthermore, the length of the correlator has to be sufficient in order to ensure small error probabilities, which is evident from (5.57).

5.4.6 Audio watermarking via the patchwork technique

The patchwork technique first presented by Bender et al. [8] for embedding watermarks in images is a statistical method based on hypothesis testing. These methods use stochastic models relying on large sets, which makes them applicabable for CD-quality audio data due to the large amount of samples.[17] The watermark encoding procedure uses a pseudorandom process to embed a certain statistic into a data set which is detected in the reading process with the help of numerical indexes (like the mean) describing the specific distribution. This method was applied to the magnitudes in the Fourier domain [28–30] in order to spread the watermark in the time domain and be more robust against random sample cropping operations.

The selection of the two subsets (see Section 2.3.3) can be described by a permutation of the indices $i = (1, \ldots, 2N)$ according to the bit to be embedded:

$$\pi = (a_1, \ldots, a_N, b_1, \ldots, b_N), \text{ with } \mathbf{pn}[a_i] = +1, \mathbf{pn}[b_i] = -1 \qquad (5.58)$$

Therefore, the watermarked block is obtained by

$$\mathbf{C_w}[n] = \mathbf{C_o}[n] + \Delta\mathbf{C_o}[n]\mathbf{pn}[n], n = \pi[i], i = 1, \ldots, 2N \qquad (5.59)$$

$$\mathbf{C_w} = \mathbf{C_o} + \mathbf{W} \qquad (5.60)$$

where the alteration of the different Fourier magnitudes is described by the vector $\Delta\mathbf{C_o}$. According to Section 2.3.3, the test performed during detection in the patchwork algorithm is a difference of subsets defined by the indexes

---

17. Every second carries 88,200 samples.

$a_1, \ldots, a_N$ and $b_1, \ldots, b_N$, which can be written as

$$\frac{1}{2N} \sum_{i=1}^{N} (\mathbf{C_w}[a_i] - \mathbf{C_w}[b_i]) = \frac{1}{2N} \sum_{i=1}^{N} \mathbf{C_w}[\pi[i]] \mathbf{pn}[\pi[i]] = \frac{1}{2} C_\tau(\mathbf{C_w}, \mathbf{pn})$$

$$(5.61)$$

Therefore, the patchwork technique in this form is equivalent to the linear correlation comparator function in the spread-spectrum technique (see above).

5.4.7 Mechanisms against desynchronization manipulations

As already mentioned in the description of several audio watermarking algorithms, the audio stream has to be synchronized to provide a successful decoding of the watermark. Particularly, the widely used statistical methods like the spread-spectrum and patchwork techniques are vulnerable against misalignment of the analyzed and watermarked blocks. In principle, an alignment is possible using a brute force method of testing all possible combinations of time shift and scaling parameters. Nevertheless, this is computationally infeasible, especially in monitoring applications (see Section 3.5) requiring real-time detection. Therefore, the synchronization algorithm integrated in the watermarking algorithms should be robust against desynchronization attacks like shifting and time- and frequency-scaling modifications, and should be fast enough. Different approaches have been developed over the last few years in order to cope with the synchronization problem and will be presented in the subsequent paragraphs.

5.4.7.1 Redundant embedding

Since spread-spectrum methods are the most widely used algorithms to watermark audio data, special attention was devoted to develop robust algorithms with respect to desynchronization attacks. If the only attack is a shift in time (due to filtering processes), synchronization can be achieved by applying sliding correlators to establish initial synchronization. After synchronization is established, the decoder enters into the mode where the watermark bits are decoded (see [10]). Nevertheless, scaling operations in time and frequency are very effective in the desynchronization of the correlator [see (5.50)] used in these algorithms. One approach to cope with the desynchronization is to embed the watermark signal redundantly [26, 31] in the time-frequency plane (see Figure 5.11) in order to reduce the number of necessary search steps that have to be performed to achieve reliable detection results.

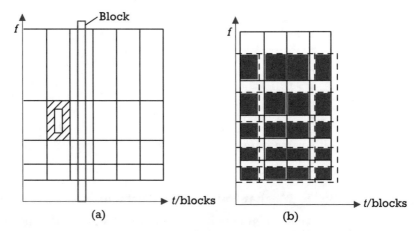

Figure 5.11 Redundant embedding in time frequency plane: (a) embedding and detection regions; and (b) maximizing overlap of watermarked and manipulated regions.

A region in the plane consists of a time segment due to several overlapping blocks and a frequency band. A number of regions are used in order to embed 1 bit of information by altering the magnitudes of the Fourier coefficients in the regions according to a pseudorandom array generated by means of a secret key. Detection is done by correlating the regions with the same pseudorandom array. In order to achieve a high overlap between the original pseudorandom array and the watermarked and manipulated time-frequency plane, the regions have to be shaped carefully. Due to the scaling property of the Fourier transformation [32]

$$\mathcal{F}[\mathbf{c_w}(kt)] = \frac{1}{|k|}\mathbf{C_w}\left(\frac{f}{k}\right) \qquad (5.62)$$

both algorithms use nonlinear frequency bands (see Figure 5.11) in order to cope with a time scale modification. Furthermore, several regions are used to embed 1 bit of information. Robustness against static time and pitch scaling is achieved in the method presented by Kirovski and Malvar [26] by performing multiple correlation tests using different combinations of time and pitch scales. Furthermore, only the center of the regions [see Figure 5.11(a)] are used for correlation which maintains the correspondence with the original marked regions in the specified scale limits. The algorithm presented by Tachibana et al. [31] uses a synchronization mechanism that performs an alignment of the watermarked and manipulated regions with regard to the maximum overlap of all the regions [see Figure 5.11(b)]. The main disadvantage of redundant embedding is the low covert channel bit rate $\approx 0.5 - 2$ bps. This is due to the fact that either the overlap has to exceed a certain value [31] or the number of samples used in the correlation process has to be high enough [26] in order to achieve a sufficiently low false probability [see (5.57)]. This again demonstrates the general tradeoff of increasing robustness at the cost of a lower bit rate.

5.4.7.2 Invariant approaches

Besides using additional preprocessing steps to provide robustness against desynchronization attacks, a preferable approach would be to develop algorithms that are inherently robust against such kinds of attacks. A method that is a generalization of the echo hiding method (see Section 5.4.3) embedding a time-shifted version of the original signal is the so-called *replica modulation* presented by Petrovic [33]. The general principle is to use a replica denoted by $r(t)$ of the cover signal itself as a carrier for the modulating signal. The replica can be generated by taking a portion of the original signal within a specified time or frequency domain and introducing a slight modification by frequency, phase, or amplitude shifting according to the secret key. The modulation of the information is performed by multiplication of the replica with a binary data signal carrying the information of the different bits

$$w(t) = \alpha m(t) r(t) \quad \text{with} \quad m(t) = \sum_{j=1}^{N} \mathbf{b}[j] h(t - jT) \tag{5.63}$$

where the sign $\mathbf{b}[j] \in \{-1, 1\}$ is determined by the jth message bit $\mathbf{m}[j] \in \{0, 1\}$ and the gain factor α is used to ensure inaudibility of the embedded watermark. During the detection, the replica is generated in the same manner as in the embedding process in order to obtain

$$\hat{r}(t) = r(t) + \alpha R[m(t)r(t)] \approx r(t) \tag{5.64}$$

where the second term represents the replica of the cover replica which should be small in comparison to $r(t)$ because of α. This signal is correlated with the watermarked signal

$$\mathbf{C}_\tau (c_w(t), \hat{r}(t)) \approx c_o(t) r(t) + \alpha m(t) r^2(t) \tag{5.65}$$

where the replica generation process has to be designed in order to minimize the first term. To decode the bits, the scaled auxiliary signal [see (5.65)] is extracted by a matched filtering corresponding to the spectrum of the auxiliary signal. If time and pitch scaling manipulations are performed, these transformations occur in the watermarked and the replica part generated from this watermarked signal simultaneously. Therefore, immunity against these desynchronization attacks is assumed if the correlation is invariant against this type of manipulation applied to both signals.

Another approach that is designed to be invariant against linear time scaling relies on the quantization index modulation presented by Mansour and Tewfik [34]. The basic idea is to modify the time interval between selected signal extremes to embed the bits of the watermark. According to the quantization step size Δ_i calculated, the interval length δ_i is adjusted to force the quantization index to be either odd ($\mathbf{m}[i] = 0$) or even ($\mathbf{m}[i] = 1$). In the detection process, the embedded bits of the watermark are computed via

$$\mathbf{m}[i] = \text{rem}\left(\left\lfloor \frac{\delta_i}{\Delta_i} \right\rfloor, 2 \right), \forall i \tag{5.66}$$

Invariance against time scaling attacks can be achieved if the quantization step size Δ_i scales in the same manner as the interval length δ_i. Therefore, the successful decoding of the bits in (5.66) relies on the exact determination of the signal extremes to obtain the correct interval length δ_i and the robustness of the calculation of the quantization step size. The salient points used are the locations of the signal edges calculated from the signal envelope, where an edge is supposed to represent a significant transition from silence to activity. In the embedding process, the time interval is adjusted in order to be in the middle of a quantization slot according to the bit to be embedded. Shifting to the middle of a quantization slot defines the correct quantization step size in the detection step which minimizes the quantization error $e(\Delta)$:

$$e(\Delta) = \min_{\Delta} \left[\frac{1}{\Delta} \sum_{i=1}^{N} |\delta_i - Q(\delta_i)| \right], \quad \text{with} \quad Q(\delta) = \Delta\left(\left\lfloor \frac{\delta}{\Delta} \right\rfloor + 0.5 \right) \tag{5.67}$$

where N is the number of intervals and $Q(\delta)$ the quantization function. The major disadvantage of this method is the low data rate, which is about 1 to 2 bps. Since this method relies on the extraction of extremes it can also be grouped into the next section using salient points to achieve robustness.

5.4.7.3 Salient-point extraction methods

Another approach to developing robust algorithms is to adjust the embedding algorithms according to salient points extracted from the audio stream. The approach presented by Wu et al. [35] aims at a special attack where samples are randomly cropped from the audio stream in order to misalign the detector without affecting the quality. The general idea is to perform embedding relative to these salient points in order to be able to resynchronize to these locations in the detection step, avoiding an exhaustive search mechanism. Clearly, a workable solution has to have a salient-point extraction procedure that is robust against attacks leading to the same locations during encoding and decoding the watermark. Therefore, this approach shifts the detecting watermarks in manipulated audio streams to the problem of computing robust features of the audio stream. The algorithm presented in [35] uses maxima in the variation of energy of the filtered original signal. To compute this variation, a ratio of two energy values computed for a block of N samples relative to the actual position is calculated:

$$\text{ratio}(i) = \frac{E_{\text{after}}(i)}{E_{\text{before}}(i)}, \quad \begin{aligned} E_{\text{before}}(i) &= \sum_{k=-N}^{-1} \mathbf{c_o^2}[i+k] \\ E_{\text{after}}(i) &= \sum_{k=0}^{N-1} \mathbf{c_o^2}[i+k] \end{aligned} \tag{5.68}$$

In principle, this algorithm can be integrated in other algorithms as a preprocessing step to provide robustness against the random sample cropping attack. However, it does not explicitly address the time scaling manipulations. Nevertheless, the information about moved salient points may be used to determine possible linear time scaling manipulation.

5.4.7.4 Embedding synchronization signals

Besides the extraction of salient points from the audio signal, early approaches have already tried to embed synchronization signals into the audio stream [36]. The idea is to detect the pilot watermark signal in order to approximate the transformation of the sampling grid that can be used for resynchronization. This idea was applied by Bäuml et al. [37] in their quantization-based audio watermarking scheme. Disadvantages of these methods are the embedding of an additional signal, possibly degrading the audio quality. Moreover, in principle, the problem of robust watermark detection is shifted to the problem of robust detection of the synchronization signal, which may be difficult without prior knowledge of the attack performed.

References

[1] Boney, L., A. H. Tewfik, and K. N. Hamdy, "Digital Watermarks for Audio Signals," *IEEE International Conference on Multimedia Computing and Systems,* Hiroshima, June 1996, pp. 473–480.

[2] Swanson, M. D., et al., "Robust Audio Watermarking Using Perceptual Masking," *Signal Processing,* Vol. 66, No. 3, May 1998, pp. 337–355.

[3] Zwicker, E., and H. Fastl, *Psychoacoustics: Facts and Models,* 2nd ed., Heidelberg: Springer-Verlag, 1999.

[4]   ISO/IEC Joint Technical Committee 1 Subcommittee 29 Working Group 11, *Information Technology—Coding of Moving Pictures and Associated Audio for Digital Storage Media at up to About 1.5 Mbit/s Part 3: Audio*, ISO/IEC 11172-3, 1993.

[5]   Shlien, S., "Guide to MPEG-1 Audio Standard," *IEEE Transactions on Broadcasting*, Vol. 40, December 1994, pp. 206–218.

[6]   Pan, D., "A Tutorial on MPEG/Audio Compression," *IEEE Multimedia*, Vol. 2, No. 2, 1995, pp. 60–74.

[7]   Zölzer, U., *Digitale Audiosignalverarbeitung*, 2nd ed., Wiesbaden: B. G. Teubner, 1997.

[8]   Bender, W., et al., "Techniques for Data Hiding," *IBM Systems Journal*, Vol. 35, Nos.(3 & 4), 1996, pp. 313–336.

[9]   Kuo, S.-S., et al., "Covert Audio Watermarking Using Perceptually Tuned Signal Independent Multiband Phase Modulation, *IEEE International Conference on Acoustics, Speech, and Signal Processing (ICASSP)*, Vol. 2, May 2002, pp. 1753–1756.

[10] Proakis, J. G., and D. M. Manolakis, *Digital Signal Processing: Principles, Algorithms and Applications*, 2nd ed., Basingstoke, U.K.: Macmillan Publishing Company, 1992.

[11] Gruhl, D., A. Lu, and W. Bender, "Echo Hiding," in R. J. Anderson, ed., *Information Hiding: First International Workshop*, Vol. 1174 of *Lecture Notes in Computer Science*, Cambridge, U.K.: Springer-Verlag, May 1996, pp. 295–315.

[12] Oh, H. O., et al., "New Echo Embedding Technique for Robust and Imperceptible Audio Watermarking," *International Conference on Acoustics, Speech and Signal Processing (ICASSP)*, Orlando, 2001, pp. 1341–1344.

[13] Ko, B.-S., R. Nishimura, and Y. Suzuki, "Time-Spread Echo Method for Digital Audio Watermarking Using PN Sequences," *International Conference on Acoustics, Speech and Signal Processing (ICASSP)*, Orlando, May 2002, pp. 2001–2004.

[14] Craver, S. A., et al., "Reading Between the Lines: Lessons from the SDMI Challenge," *Proceedings of the 10th USENIX Security Symposium*, Washington D.C., August 2001.

[15] Petrovic, R., et al., "Data Hiding Within Audio Signals," *4th International Conference on Telecommunications in Modern Satellite, Cable and Broadcasting Service*, Nis, Yugoslavia, October 1999, pp. 88–95.

[16] Petitcolas, F. A. P., R. J. Anderson, and M. G. Kuhn, "Attacks on Copyright Marking Systems," in D. Aucsmith, (ed.), *Information Hiding: Second International Workshop*, Vol. 1525 of *Lecture Notes in Computer Science*, Portland, OR: Springer-Verlag, April 1998, pp. 218–238.

[17] Nahrstedt, K., and L. Qiao, "Non-Invertible Watermarking Methods for MPEG Video and Audio," in J. Dittmann et al., (eds.), *Multimedia and Security Workshop at ACM Multimedia 98*, Bristol, U.K.: September 1998, pp. 93–98.

[18] Lacy, J., et al., "Intellectual Property Protection Systems and Digital Watermarking," *Optics Express*, Vol. 3, No. 12, December 1998, pp. 478–484.

[19] Dittmann, J., "Sicherheit in Medienströmen: Digitale Wasserzeichen," Ph.D. thesis, TU Darmstadt, 1999.

[20] Cheng, S., H. Yu, and Z. Xiong, "Enhanced Spread Spectrum Watermarking of MPEG-2 AAC Audio," *IEEE International Conference on Acoustics, Speech, and Signal Processing (ICASSP)*, Vol. 4, May 2002, pp. 3728–3731.

[21] Bosi, M., et al., "ISO/IEC MPEG-2 Advanced Audio Coding," *Journal of the Audio Engineering Society*, Vol. 45, No. 10, October 1997, pp. 789–814.

[22] ISO/IEC Joint Technical Committee 1 Subcommittee 29 Working Group 11, "Coding of Moving Pictures and Audio," *MPEG-2 Advanced Audio Coding, AAC*, 1997.

[23] Neubauer, C., "Digitale Wasserzeichen für unkomprimierte und komprimierte Audiodaten," in R. Steinmetz and M. Schumacher, (eds.), *Sicherheit in Netzen und Medienströmen*, Heidelberg: Springer-Verlag, September 2000, pp. 149–158.

[24] Haitsma, J., et al., "Audio Watermarking for Monitoring and Copy Protection," *Proceedings of the ACM Multimedia 2000 Workshop*, Los Angeles, November 2000, pp. 119–122.

[25] Kirovski, D., and H. Malvar, "Robust Spread-Spectrum Audio Watermarking," *IEEE International Conference on Acoustics, Speech and Signal Processing (ICASSP)*, Salt Lake City, May 2001, pp. 1345–1348.

[26] Kirovski, D., and H. Malvar, "Robust Covert Communication over a Public Audio Channel Using Spread Spectrum," in I. S. Moskowitz, (ed.), *Information Hiding: 4th International Workshop*, Vol. 2137 of *Lecture Notes in Computer Science*, Portland, OR: Springer-Verlag, April 2001, pp. 354–368.

[27] Bassia, P., and I. Pitas, "Robust Audio Watermarking in the Time Domain," in S. Theodoridis et al., (eds.), *Signal Processing IX, Theories and Applications: Proceedings of EUSIPCO-98, Ninth European Signal Processing Conference*, Typorama Editions, September 1998, pp. 25–28.

[28] Arnold, M., and S. Kanka, "MP3 Robust Audio Watermarking," *DFG V III D II Watermarking Workshop*, Erlangen, Germany, October 1999.

[29] Arnold, M., "Audio Watermarking: Features, Applications and Algorithms," *Proceedings of the IEEE International Conference on Multimedia and Expo (ICME 2000)*, New York, July 2000, pp. 1013–1016.

[30] Yeo, I.-K., and H. J. Kim, "Modified Patchwork Algorithm: A Novel Audio Watermarking Scheme," *International Conference on Information Technology: Coding and Computing*, Las Vegas, April 2000, pp. 237–242.

[31] Tachibana, R., et al., "An Audio Watermarking Method Robust Against Time- and Frequency-Fluctuation," in P. W. Wong and E. J. Delp, (eds.), *Proceedings*

of Electronic Imaging 2001, Security and Watermarking of Multimedia Contents, Vol. 4314, San Jose, CA, 2001, pp. 104–115.

[32] Bracewell, R. N., *The Fourier Transform and Its Applications*, New York: McGraw-Hill, 1986.

[33] Petrovic, R., "Audio Signal Watermarking Based on Replica Modulation," *5th International Conference on Telecommunications in Modern Satellite, Cable and Broadcasting Service*, Nis, Yugoslavia, September 2001, pp. 227–234.

[34] Mansour, M. F., and A. H. Tewfik, "Audio Watermarking by Time-Scale Modification," *IEEE International Conference on Acoustics, Speech and Signal Processing (ICASSP)*, Salt Lake City, May 2001, pp. 1353–1356.

[35] Wu, C. P., P. C. Su, and C.-C. J. Kuo, "Robust and Efficient Digital Audio Watermarking Using Audio Content Analysis," in P. W. Wong and E. J. Delp, (eds.), *Proceedings of Electronic Imaging 2000, Security and Watermarking of Multimedia Contents*, San Jose, CA, January 2000, pp. 382–392.

[36] Tilki, J. F., and A. A. L. Beex, "Encoding a Hidden Auxiliary Channel onto a Sigital Audio Signal Using Psychoacoustic Masking," *IEEE Souteastcon*, April 1997, pp. 331–333.

[37] Bäuml, R., (et al.), "A Channel Model for Watermarks Subject to Desynchronization Attacks," in P. W. Wong and E. J. Delp, (eds.), *Proceedings of Electronic Imaging 2002, Security and Watermarking of Multimedia Contents IV*, Vol. 4675, San Jose, CA, January 2002, pp. 281–292.

Contents

Digital watermarking for other media

In this chapter the main watermarking principles for the media types video, three-dimensional data, two-dimensional geometry data, text, and printed music scores are described. Because of current developments, which are elaborated in Section 6.2, increased significance of three-dimensional data may be assumed in the future. Thus the focus in this chapter is on three-dimensional watermarking principles.

6.1 Digital watermarking for video data

While the majority of research in the field of digital watermarking techniques has apparently been conducted on still images, other media types have also been paid attention to by the research community. These particularly include video data. The increasing bandwidth of Internet connections already allows the transmission of video streams of publicly accepted quality: The distribution of video material equivalent in quality to analog home video is available for users with broadband connections willing to wait for a download to complete within several minutes to hours. In addition, digital movies are already broadcast to cinemas, while DVD video is rapidly replacing traditional analog videotapes for rental and buy-to-own video.

6.1.1 Application requirements

Watermarking video data can be relevant in several scenarios. Among these scenarios are proof of ownership, broadcast or distribution monitoring, integrity checking (e.g., in video surveillance systems), authentication (e.g., for identification of a video source), and usage control (to limit copy operations). Each of these scenarios has its individual requirements. Currently, the most relevant scenario is the *production and broadcasting scenario* of the European Broadcasting Union (EBU). This scenario is the basis for a video watermarking evaluation that was performed by the EBU in 2001 [1].

The requirements of the EBU scenario can be summarized as follows: Even high-quality video source material intended for further processing is watermarked. Therefore, the quality of the watermarked data must be very high. The embedded watermarks must be invisible even at high studio quality. Embedding a payload of 64 bits is usually considered in this scenario. However, the quality requirement and the related watermark minimum segment (WMS: the minimum video segment, which must contain a watermark) are challenging for watermark developers. Ideally, the embedded watermark should always be detected in watermarked material. Thus, the detection probability should be very high if video data are watermarked. Conversely, no watermark should be found in unwatermarked material, which is equivalent to a very low false detection probability. This very low false detection probability is important if millions of devices with watermark detectors are in use worldwide. Video data are processed before being distributed to the end consumer. Concerning robustness, a watermark must survive these processing operations which include compression (for the distribution of video on channels with capacity constraints) and format conversion (to address the different national television standards). Embedding and detection must be possible in real time, which is essential for monitoring. However, a small latency cannot be avoided. Blind detection is required: Nonblind watermarking systems in the monitoring scenario require the use of multiple monitoring systems, one system for each single video material multiplied by the number of channels that should be watermarked. Robustness against targeted attacks must be considered in some scenarios. The techniques applied by attackers can range from simple attacks like geometric transformations and frame dropping to rather complex protocol schemes.

The EBU restricted attacks in their scenario, which is shown in Figure 6.1, to so-called *production attacks*. This scenario consists of three parts:

1. *Production level:* High-quality material is used during the production of a video. The data rate is approximately 50 Mbps. Requirements for the watermark mainly include invisibility and robustness, especially to the processing steps, the so-called *production attacks*. Typical operations are scaling, rotation, format conversion, color space conversion,

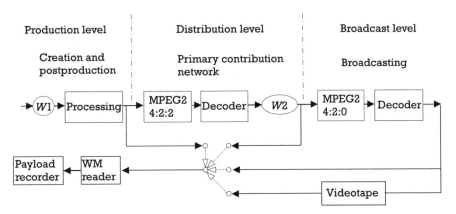

Figure 6.1 The EBU scenario consists of three parts: production, distribution, and broadcast levels. Watermarks are embedded in the production and distribution levels. Attacks are limited to typical processing operations during the production.

frame rate conversion, and compression. The WMS for high-quality video is 1 sec.

2. *Distribution level:* After postprocessing the video is compressed. The data rate ranges from 8 to 20 Mbps. The WMS is 5 sec for medium quality.

3. *Broadcast level:* Before broadcasting, the video will be compressed again to meet the channel capacity limits.

All watermarks embedded have to store 64 bits in each WMS. The quality of the watermarked video data is evaluated by subjective tests (see benchmarks in Section 7.3.1).

6.1.2 Algorithms

Similarly to audio watermarking methods, video watermarking methods can be distinguished by the temporal order of embedding in relation to compression. While embedding before compression allows one to choose the watermarking methods arbitrarily, embedding during compression is limited to the transformation domain of the compression algorithm. Embedding after compression modifies the compressed video bit stream. Typically for all these schemes, a requirement is that watermark retrieval is possible from uncompressed video streams.

As in image watermarking, a number of different schemes have been proposed for video watermarking. However, video watermarking is not just the application of an image watermarking technique to each individual frame. Temporal dependencies provide additional difficulties, but also additional opportunities.

Embedding in the spatial domain is one of the characteristics of the JAWS [2] video watermarking algorithm by Kalker et al.; JAWS is the acronym for Just Another Watermarking System. It embeds a watermark pattern **w** in the spatial domain by changing intensity values to guarantee a robustness against color conversions. A spatial correlation is used to decide whether a watermark is embedded or not: If the correlation value C_τ exceeds a certain threshold τ, the watermark is detected. This allows the embedding of an 1-bit payload.

Because of complexity reasons, utilizing the temporal axis was not considered by Kalker et al. Instead, the watermark pattern $\mathbf{w} = \mathbf{w}[i]$ is repeatedly embedded in every frame $\mathbf{c_o} = \{\mathbf{c_o}[i]\}$. The index i denotes the pixel position. The watermark **w** is an additive watermark and is weighted using a global scaling factor α, which affects the embedding strength. The watermark consists of independent normal distributed floating point values with $\mu = 0$ and $\sigma^2 = 1$. Using other than spectrally white patterns would increase the robustness of the watermark, but would also result in more visible artifacts. Still artifacts are visible in regions containing no motion and weak textures. Therefore, a local scaling factor $\boldsymbol{\lambda} = \{\lambda[i]\}$ is introduced. A highpass filter[1] is used to determine $\lambda[i]$. The embedding process can be described as

$$\mathbf{c_w}[i] = \mathbf{c_o}[i] + \alpha\lambda[i]\mathbf{w}[i] \tag{6.1}$$

To incorporate shift invariance, a watermark pattern is tiled and, if necessary, truncated over one frame, which permits an improved exhaustive search over all possible shift values. This is shown in Figure 6.2.

The repetition allows the frame data $\mathbf{c_w}$ to be folded[2] before calculating the correlation matrix. The correlation value can be calculated efficiently by using a cyclic convolution. Kalker et al. extended JAWS to embed an increased payload by using different patterns which store the information in the relative positions of the correlation peaks.

---

1. A highpass filter detects edges and results in low values for uniform frame regions.
2. A folded version of a frame $\mathbf{c_w}$ is obtained by partitioning $\mathbf{c_w}$ in adjacent blocks $\mathbf{c_{wi}}$ with the size of a single watermark tile followed by a pointwise summation:

$$\text{fold}(\mathbf{c_w}) = \sum_i \mathbf{c_{wi}} \tag{6.2}$$

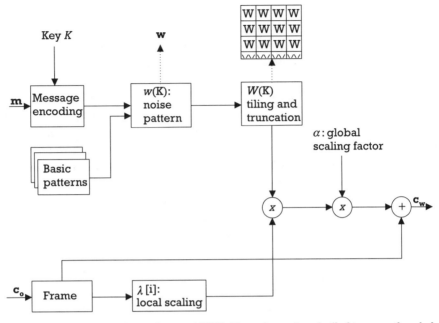

Figure 6.2 The embedding scheme of JAWS. The noise pattern is tiled to cover the whole frame. The embedding strength is determined by a global and a local scaling factor.

Embedding in the transformation domain can be similar to image watermarking in the transformation domain as realized in the SysCoP video watermarking algorithm by Busch et al. [3]. Busch et al. make some modifications to the original image watermarking algorithm (see Section 4.2.1 and [4]). Real-time-capable implementations of the DCT and inverse DCT are used. The SysCoP algorithm is executed on a digital signal processor (DSP) board. Visual quality is increased by checking if the block that is used for watermarking contains textures or edges. To reduce the complexity of the algorithm, this is done by evaluating the DCT coefficients. The low-frequency DCT coefficients contain information about the existence of edges or plain areas in the block. Blocks identified as plain areas are watermarked with lower strength. Blocks identified as edge-containing areas are skipped. Robustness against MPEG 2 compression is achieved by maximum redundancy. Almost all blocks of a video frame are subjected to the watermarking procedure. Shift resistance is realized by detecting the watermarked frame's origin.

However, as discussed before, video sequences allow a wider range of attacks compared to images because of their additional temporal dimension.

By removing some individual frames from video sequences, which is called *frame dropping*, the visual quality will not suffer in frames containing no or mainly slow motion. Conversely, frames can be duplicated, interpolated, or exchanged. Averaging attacks build an average frame based on some consecutive frames which can be applied to static frame regions. One possible countermeasure is the embedding of the same watermark in static regions. Yet dynamic frame regions must be treated differently: If the same watermark is embedded in regions with motion, an attacker can estimate similarities in these regions and try to remove them. Averaging and collusion-based attacks cannot be neglected because of the high redundancy of video material. Su et al. [5] addressed the issue of collusion-resistant video watermarking. Further video-specific attacks are described in detail by Deguillaume et al. [6]. Intentional attacks are challenging for video watermarking developers even if one watermark content is embedded in every frame.

A watermarking algorithm that embeds the watermark and a synchronization template in the time domain was presented by Niu et al. [7]. This algorithm provides robustness against rotation, scaling, and translation.

Another approach based on structural noise patterns was presented by Lagendijk et al. [8]. Instead of reversing the distortion, this algorithm embeds a pattern that is key dependent. For retrieving the watermark, a key is used to create a custom filter. Only this key-dependent custom filter is assumed to be able to reveal the embedded pattern. However, the problem of complex geometric and estimation-based attacks are not addressed in this work.

Embedding in the compressed domain is addressed by Jordan et al. [9]. They propose modifications of the motion vectors to embed the watermark information. While the embedded information can be retrieved from compressed video sequences, retrieving the embedded information from decompressed sequences is unlikely. Hartung and Girod [10] proposed a method that is capable of embedding the information also in the compressed domain and retrieving the information from the decompressed domain. The general scheme of the proposed method is shown in Figure 6.3.

Only the DCT coefficients of the MPEG 2 bit stream are modified. Before modification, MPEG 2 compression operations (entropy encoding and quantization) are inverted. Thus, this scheme is in fact embedding in the transformation domain. After adding the watermark, which is a two-dimensional signal and was transformed in the DCT domain, the new bit rate is compared with the original and, depending on the bit rate, the original DCT block is selected. This allows controlling the bit rate, but influences redundancy and robustness.

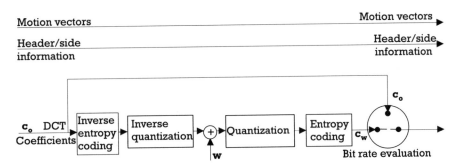

Figure 6.3 Block diagram of a compressed video watermarking scheme as proposed by Hartung and Girod [10]. After the watermark is embedded, the bit rate is evalutated. If it exceeds a certain limit, the watermarked DCT coefficients are replaced by the original ones.

6.2 Digital watermarking for three-dimensional data

The processing capabilities of current computer hardware easily permit handling of three-dimensional data on desktop computers with continuous rapid performance improvements: Near future architectures will allow desktop computer graphics to achieve cinematic results in real time. With ongoing development in this field, the available computational capabilities will lessen the differences between computer-generated and man-made video data drastically. This assessment is also confirmed by the fact that graphical processor units (GPU) are used in areas that were out of scope during their development. This appears to validate the assumption that more and more three-dimensional data will be used and exchanged, even though the current use and exchange of three-dimensional data are still in their infancy. Because of this development, watermarking of three-dimensional data may be assumed to become increasingly important. The extended future usage of three-dimensional models is already indicated by the so-called synthetic natural hybrid coding (SNHC) of the MPEG 4 video compression scheme. *Synthetic* images (like three-dimensional computer animations) complement *natural* audio and video data.

6.2.1 Application requirements

Various applications can be identified in which three-dimensional data are used. Each of these applications has different demands on the underlying data. Manufacturing processes are based on software like computer-aided design (CAD) for structural design or finite element method (FEM) for physical modeling of products. The results are real-world products. Therefore, accuracy and high precision are necessary to guarantee a functioning product.

Even the slightest modification of data can lead to unwanted effects or even destroy the intended functionality. On the other hand, animations of movie sequences, virtual worlds, and computer games, or ray tracing are not as strongly dependent on the accuracy of data as long as the results are visually appealing and correspond to real-life experiences. Obviously, the quality of animated sequences can also be affected by changes of data that might, for example, result in gaps or intersections that were not part of the original scenario. Currently, the demands on virtual worlds or computer games are not as high as for animated movie sequences. However, as mentioned before, their quality is rapidly increasing due to current developments, particularly emanating from the computer gaming sector. Therefore, it appears safe to predict that future demands will also increase. Medical applications have unique demands, too. Usually, changes of medical data are not accepted by the medical profession. The main concern here is that a misdiagnosis must not be caused by the changes introduced by watermarking techniques. This problem is also encountered in the application of image compression techniques in medical application scenarios. While voxels (see Section 6.2.2) are common in computed tomography (CT) or in magnetic resonance imaging (MRI), CAD models are used for designing prostheses. These CAD models underlie the above-mentioned manufacturing requirements.

6.2.2 Data representation

As with other media types, type-specific algorithms must be developed for three-dimensional data; this section provides a brief introduction to the underlying representations.

6.2.2.1 Voxels

Two-dimensional images are referred to as an image grid consisting of individual picture elements (pixels). A stack of two-dimensional images as shown in Figure 6.4 can be considered as three-dimensional data consisting of volume elements (voxels).

2d-image lattice: $\{(x, y) \mid 0 \leq x < N_{columns}, 0 \leq y < N_{rows}\}$

3d-voxel lattice: $\{(x, y, z) \mid 0 \leq x < N_{columns}, 0 \leq y < N_{rows}, 0 \leq z < N_{slices}\}$

$$(6.3)$$

Similar neighborhoods, filter masks, and operations can be defined as described by Lohman [11] analogously to images.

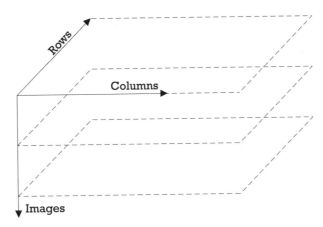

Figure 6.4   Voxels can be regarded as a stack of two-dimensional images.

6.2.2.2   Solids

Different voxels can be considered as belonging to one solid. In constructive solid geometry (CSG) [12–14], basic primitives are, for example, spheres, cones, cylinders, or rectangular solids. Boolean operations and linear transforms can be used to design more complex models. This is typically used by engineers for modeling volumetric objects, such as tubes, or by ray-tracing engines. A complex object is represented by a CSG tree which reflects the construction from primitives.

6.2.2.3   Patches

Patches represent the exact surfaces of objects. Usually these are polynomials in two parametric variables and cubic. A pth-degree nonuniform rational B-spline (NURBS) curve **C** defines a trajectory in a three-dimensional space by a set of control points and their weights. The weights allow a subtle control of the NURBS curve, which is different from moving a control point.

NURBS surfaces can be defined in a compact and efficient representation. Detailed introductions can be found in [14–16].

6.2.2.4   Meshes

The surface of objects can be approximated by a mesh of planar polygonal facets. These are called *boundary representations* or *simplicial complexes*. Meshes are commonly used representations of three-dimensional objects.

$$\mathcal{M} = (\mathcal{E}, \mathcal{V}) \quad \text{where } \mathcal{E} = \{e[0], \dots, e[m]\}: \quad \text{connectivity of edges} \\ \text{and} \quad \mathcal{V} = \{\mathbf{v}[1], \dots, \mathbf{v}[n]\}: \quad \text{set of vertices} \tag{6.4}$$

Additional attributes like color or texture coordinates are possible in this representation.

The topology of meshes can be characterized as follows: A mesh can consist of triangles, quadrangles, or in general of polygons with N edges. It can be open or closed, free of self-intersections or self-intersecting, or can have different manifolds, which is related to the number of faces adjacent to a single edge.

Topological representations can be constructed by using multiresolution techniques. Algorithms of the previous class decompose a given geometry into its basic shape and detail information on different levels of resolution. The theoretical background of multiresolution transforms is explained in [17]. Further information on mesh decimation and simplification can be found in [18].

6.2.2.5 Point clouds

Another representation of three-dimensional models are point clouds. Point clouds consist of a set of three-dimensional points. These point clouds can be sampled from the surface of existing models, for example, by laser scanners. Further processing steps convert the point clouds to surfaces. Hoppe et al. [19] first stated this abstract problem.

6.2.3 Operations and attacks

Typical operations and requirements are related to the data representation; the following provides an outline of mesh data and voxel operations.

Mesh data   Benedens [20] proposed required robustness properties of three-dimensional watermarking systems. Concerning robustness, he grouped relevant operations according to the application scenarios.

Unintentional attacks on labeling provide rather restricted scenarios. Randomization or quantization of coordinates is possible due to reduction of the data representation. Randomization of vertex or face order or of vertices in a face might occur due to internal representation differences in available graphics software packages. Additional attacks are caused by creating a new scene graph by combining existing data, which may involve Euclidean transformations, uniform scaling, general affine transformations, format conversions, removal of parts of the mesh, or the joining of a mesh with others.

Unintentional attacks on proof of ownership or on fingerprinting result in a more severe impact on the design of watermarking algorithms. Modifications that are not limited to simple operations for the creation of a scene graph must not affect watermarks in this application scenario. Some of the operations are surface smoothing, multiresolution editing, signal processing, free-form deformations, and subdivision surfaces. These operations are typically applied to adapt objects to specific scenes. Further noise is added by compression or conversion between different representations. Additionally, three-dimensional models can also be *printed* (e.g., they can be plotted by using stereo lithography). Currently, this is only important for rapid prototyping. But the increasing number of available milling and plotting machines might lead to comparable consequences well-known from photocopiers.

Intentional attacks are, for example, the use of compression techniques like polygon reduction which can be followed by methods that smooth the reduced meshes (e.g., subdivision surfaces). Attacks well known from the image watermarking area, like embedding multiple watermarks with the same scheme, synchronization attacks, watermark estimation, and removal or protocol attacks can be transferred to three-dimensional data representations.

Voxels represent content that must not be changed. This affected the digital imaging and communications in medicine (DICOM) format definition for volume data representation. Medical images are typically stored in uncompressed formats. Therefore, watermarking techniques for voxels and other medical images should be invertible or erasable to meet the needs of medical applications. An invertible watermarking technique was presented by Fridrich [21]. However, no invertible scheme for voxels has been presented, although Fridrich's method can be extended to three-dimensional data.

Operations and attacks can be performed either on each individual slice or directly on three-dimensional data. Operations on individual slides are related to operations on images and can be found in Chapter 4. These operations can also be extended to three-dimensional data, for example, by modifying the neighborhood definition. A number of special processing operations have been defined for voxels; details on these operations can be found in publications on medical image processing [11, 22].

6.2.4 Algorithms

Watermarking algorithms differ in the embedding primitives used.

6.2.4.1 Embedding primitives for three-dimensional geometry data

Concerning the embedding primitives, a distinction between mesh data and voxels must be made.

Mesh data provides at least information about geometry (*shape*) and topology (*vertex connections*). In some cases additional information for rendering is available.

Geometrical embedding primitives are invariant to certain transformations, providing implicit robustness. Invariants of polygonal meshes are listed in Table 6.1. Also, invariants on point sets in $\mathbb{R}^n$ exist, like the Euclidean metric ($\rho(\mathbf{p}, \mathbf{y}) = \sqrt{\sum_{i=1}^{n} |\mathbf{p}[i] - \mathbf{y}[i]|^2}$) or the Mahalanobis norm ($r(\mathbf{p}) = \sqrt{(\mathbf{p} - \overline{\mathbf{x}})^T \mathbf{C}_{\mathcal{X}}^{-1} (\mathbf{p} - \overline{\mathbf{x}})}$, where the matrix $\mathbf{C}_{\mathcal{X}}$ is defined by the set of points $\mathcal{X}$ with its center of mass $\overline{\mathbf{x}}$) or the Nielson-Foley norm described later (see Section 6.2.4). For embedding, the precision of floating point values is an important issue. For example, the IEEE single (double) precision standard provides a precision of 6 (12) decimal digits. Additional impacts can be caused by other factors like export filters of application programs.

Topological embedding primitives are robust against geometrical transformations. Information is embedded by adding, removing, or displacing (changing the adjacency of) vertices. Topological modifications like mesh decimation [18] affect the embedded information.

Other embedding primitives are given by additional information of three-dimensional models. The description of geometrical three-dimensional objects is not limited to vertices and their connectivity. Additional information is necessary to allow a natural appearance, especially in rendering scenarios. Objects can have predefined attributes like reflectivity, per-vertex colors, per-vertex normal vectors, per-vertex texture coordinates, and texture images. A watermarking method that modifies the per-vertex texture coordinates was presented in [23].

Voxels can be watermarked slice by slice with image watermarking techniques. Video watermarking techniques, which are developed for sequences of images, might be straightforward, too. However, the additional third dimension allows operations that were neither faced nor foreseen during

Table 6.1 Invariants of Polygonal Meshes

Primitives	Invariant Against
Point coordinates	—
Line length, polygon area, and polyhedron volume	Rigid transformations
Two triangles describing quantities and ratio of two polygon areas	Euclidean transformations
Ratio of two line lengths, ratio of two polyhedron volumes, and barycentric combination	Affine transformation
Cross-ratio of four points on a straight line	Perspective transformations

the development of watermarking algorithms for images and video data. For example, rotating a three-dimensional object arbitrarily in three dimensions is a common operation, whereas rotating a movie sequence is somewhat limited in application.

6.2.4.2   Overview of existing methods

Mesh watermarking methods based on topological alterations were first presented by Ohbuchi et al. [24]. He calls this method mesh density pattern (MDP) embedding. The algorithm tessellates given surfaces. An example of a water-marked surface is given in Figure 6.5.

The triangle strip peeling symbol sequence (TSPS) embedding algorithm also proposed by Ohbuchi et al. [23] changes the topology of the original mesh. A triangle strip is peeled off from the original mesh as shown in Figure 6.6.

After the selection of a starting edge e and the starting triangle, the message bits are used to determine the next triangle by using the other two free edges of the current triangle. This procedure is continued until all message bits are embedded. Additional control information is interleaved with message bits to avoid the strip hitting the boundary or circling back to itself. Embedding is related to the quantization index modulation (see Section 2.3.3). The resulting strip S is peeled off from the original mesh by splitting all the

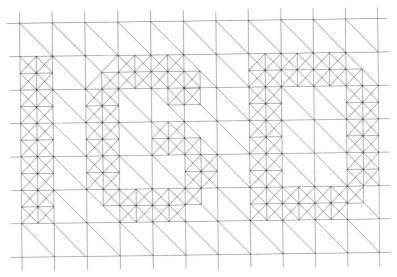

Figure 6.5 Ohbuchi's MDP method [24] embeds information visibly by changing the tessellation of a given mesh.

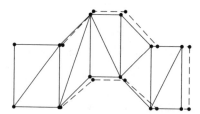

——— Peeled strip S

— — Original mesh

Figure 6.6 Triangle strip peeling symbol sequence: The strip is peeled off by adding new vertices and edges and changing the original connectivity.

edges and vertices on the boundary of S except the initial edge. An improved version of this algorithm was presented by Cayre and Macq [25] by increasing its security.

Watermarking methods, which are based on topology changes, have the advantage that geometrical operations that alter the locations of the points do not interfere with the embedded information. However, topology-changing operations like mesh simplification, which are compared in [26], are quite common in certain scenarios, limiting the applicability of this approach.

Mesh watermarking methods for polygonal models based on quantities defining a set of similar triangles modify the geometry of the model. Ohbuchi et al. presented two methods that change the geometry of models. The triangle similarity quadruple (TSQ) method of embedding [24] uses pairs of dimensionless quantities as information carriers. These pairs define a set of similar triangles. A quadruple of adjacent triangles is considered as an embedding primitive, which Ohbuchi called a macroembedding primitive (MEP). Each MEP stores a quadruple of values (as shown in Figure 6.7): {marker M, subscript S, data D_1, data D_2}.

The marker identifies the modified MEPs used for embedding. For embedding, the watermark is split into data symbols. The subscript is a rearrangement criterion for the split watermark message.

The following three steps are repeated until all data symbols are embedded:

1. The input triangular mesh is searched for a set of four triangles that can be used as an MEP. Vertices that have already been used for embedding and triangles that are not suitable for stable embedding (e.g., because of numerical instability) have to be rejected.

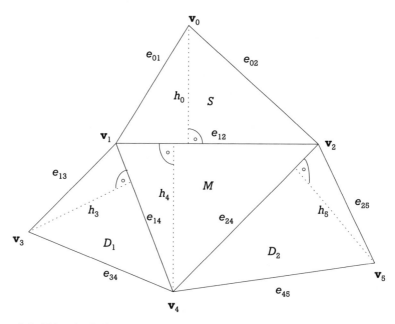

Figure 6.7   Triangle similarity quadruple: The macroembedding primitives used to store a marker M, a subscript S, and two data values D_1 and D_2. The marker is necessary to identify MEPs storing information. By using the subscript information, the original data string is assembled from the individual data values.

2. The marker value is embedded by changing the quantity pair $\{e_{14}/e_{24},\ h_4/e_{12}\}$, where $e_{ij}(h_i)$ is the length of the edge $\mathbf{v}_i\mathbf{v}_j$ (the height of the triangle). This affects the vertices $\mathbf{v}_1$, $\mathbf{v}_2$, and $\mathbf{v}_4$.

3. The subscript and two data symbols are embedded in the remaining three triangles. Because of the marker value already stored, only the vertices $\mathbf{v}_0$, $\mathbf{v}_3$, and $\mathbf{v}_5$ can be modified. First, the ratio $\frac{h_i}{e_{ij}}$ is modified by changing only h_i. Subsequently, the ratio $\frac{e_{ij}}{e_{kl}}$ is modified while keeping the height constant.

The magnitude of the changes has to be chosen carefully as a trade-off between robustness, capacity, and visibility. The watermark retrieval process consists of two steps:

1. For each located MEP, the subscript and the data symbols are extracted and stored.

2. The subscript values are used to rearrange (sort) the data symbols.

The presented implementation used a majority vote for error correction (if multiple symbols had been embedded).

In Benedens' triangle flood algorithm (TFA) [27], modifying the triangles' heights creates a unique traversal path similar to the TSPS embedding method. This traversal path will store the watermark by modifying the vertices along the path. In this approach, modified triangles are not stripped from the original mesh; mending algorithms [28] will not affect this kind of watermark.

Mesh altering methods for polygonal models based on volume ratios of pairs of tetrahedrons are the embedding primitives in the Ohbuchis tetrahedral volume ratio (TVR) scheme [24]. Providing a global one-dimensional ordering of the embedding primitives is the first step of this algorithm. These primitives are used to store one symbol in each volume ratio of neighboring triangles. The proposed algorithm is a blind watermarking technique. The information is embedded locally and robust against affine transformations. The algorithm is only applicable to two manifolds. Meshes must consist of triangles. As shown by Benedens [29], double IEEE floating point precision is necessary.

The previously described methods are based on triangles or tetrahedrons. Remeshing operations or polygon simplification will remove information embedded by these watermarking techniques. The main advantages of these algorithms are capacity and execution speed.

Mesh altering methods for polygonal models based on vertices distances were proposed by Benedens' Vertex flooding algorithm (VFA) [30] which does not require connectivity of the faces in the input mesh. This algorithm modifies the vertices' distances to the center of mass for encoding the information:

1. After selecting a start triangle, its center of mass $\bar{\mathbf{v}}$ is calculated. The edge points $S = \{s_1, s_2, s_3\}$ of the start triangle are not considered in the further steps.

2. The vertices are partitioned: $M_k = \{\mathbf{v} \in V| \quad k \leq \frac{|\mathbf{v}-\bar{\mathbf{v}}|}{W} < k+1\}$, $0 \leq k \leq \frac{r}{W}$;

3. Subpartitions are created: $N_{k,l} = \{\mathbf{v} \in M_k| \quad l \leq \frac{|\mathbf{v}-\bar{\mathbf{v}}|-kW}{2^n} < l+1\}$, $0 \leq l \leq 2^n - 1$.

In each partition M_k, one bit string with the length n is embedded. The value l is embedded by moving the vertices of M_k to the subpartition of $N_{k,l}$. For reading the watermark, the mean norm values of the partition M_k and the subpartitions $N_{k,l}$ are calculated. The index k represents the position in the bit string and l its value. The embedded watermark is a fragile watermark. It is public and hence can be used for authentication or integrity checks.

Mesh altering methods for polygonal models based on distributions of normal vectors were proposed by Benedens [30]. His normal bin encoding (NBE) algorithm was designed to survive point randomization, remeshing operations, and polygon simplifications. The algorithm can be described as follows:

1. The unit sphere is partitioned into bins. Each bin is described by its center and its radius, which is shown in Figure 6.8.

2. Each normal vector of the three-dimensional model is assigned to a bin.

3. The distribution of the differences of the normal vectors to the bin centers are calculated.

For embedding, some bins are selected using a secret key. The normal distributions in the selected bins are changed. Watermark retrieval is based on a hypothesis test: The histograms of the selected and nonselected bins are calculated. A hypothesis test checks if these two histogram distributions are identical. This watermarking method is blind. Only the key but not the original model is necessary for retrieving the watermark. It is robust against affine transformations, polygon reduction, and cropping. A reorientation of the model is necessary before the watermark retrieval process.

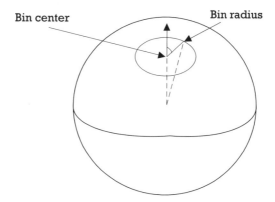

Figure 6.8 Normal bin encoding: The unit sphere is partitioned into several bins. Each bin is described by its center and its radius. The watermark is embedded by changing the distribution of the normal vectors in certain bins.

Mesh altering methods for polygonal models based on affine invariant norms were proposed by Wagner [31]. This watermarking scheme embeds a 1-bit watermark (in contrast to the NBE which can embed string values). The inertia ellipsoid of a point set, which is also called the *Nielson-Foley norm* and which was discussed in [32], is the basis for his method. First, the centroid $\bar{\mathbf{v}}$ is calculated. This centroid is used to calculate a new coordinate system defined by the matrix $\mathbf{V}$:

$$
\mathbf{V} = \begin{pmatrix} x[0] - \bar{x} & y[0] - \bar{y} & z[0] - \bar{z} \\ \vdots & \vdots & \vdots \\ x[n] - \bar{x} & y[n] - \bar{y} & z[n] - \bar{z} \end{pmatrix} \tag{6.5}
$$

The used norm is defined by

$$
\|\mathbf{v}\|_{\mathbf{P}} = v\mathbf{A}v^T = (x \quad y \quad z)\, \mathbf{A} \begin{pmatrix} x \\ y \\ z \end{pmatrix} \qquad \text{where} \quad \mathbf{A} = (n+1)(\mathbf{V}^T\mathbf{V})^{-1} \tag{6.6}
$$

This norm is applied as follows: For each vertex $\mathbf{v}[i]$, a "star" $\mathcal{S}_i$ is defined by its center vertex $\mathbf{v}[i]$ and the connected surrounding vertices. This star is used to calculate a local approximation of the surface normal in vertex $\mathbf{v}[i]$, which is affine invariantly connected with the mesh M: $\mathbf{n}[i] = \frac{1}{|\mathcal{S}_i|}\sum_{j\in\mathcal{S}_i}\mathbf{v}[i] - \mathbf{v}[j]$. Each individual normal vector length is converted into an integer value by using the average normal vector length l: $n[i] = \text{round}(\frac{c}{l}\|\mathbf{n}[i]\|)$. Here, c is an arbitrary number. For embedding information, the integer approximations of the normal vector lengths are changed. Wagner suggests a function defined over a sphere, such as a logo that was projected on a unit sphere, as message. Aspert et al. [33] extends this algorithm to arbitrary embedding strings. This function is discretized at the positions of the normal vectors using a given resolution (b is the number of bits used to store the information):

$$
w[i] = \text{round}\left(2^b f\left(\frac{\mathbf{n}[i]}{\|\mathbf{n}[i]\|}\right)\right) \tag{6.7}
$$

The bits of the discretized positions are embedded in the magnitude of the normal vectors. The new normal magnitudes are used for recalculation of the vertex positions. This is done by solving a system of linear equations. Therefore, the changes result in deformations of stars.

Using the Nielson-Foley norm was also proposed by Benedens [29]. Instead of embedding the information globally, the information is embedded locally in this approach, which can be compared with Ohbuchi's TVR and TSQ methods discussed before. The information is also embedded in the ratios of two tetrahedrons (like TVR). The possible value range is subdivided into intervals. The connected vertices are modified to fall into certain intervals. One embedding primitive also stores bit values with their corresponding position in the watermarking string (like TSQ). While this method is robust against affine transformations, it is not robust against remeshing.

Mesh altering methods for polygonal models based on multiresolution representation have been proposed that differ in the used representation and the way of embedding information. Kanai et al. [34] propose a method that is based on the wavelet transform and a multiresolution representation resulting from the wavelet decomposition, specifically the lazy wavelet transform [35]. This transform decomposes the high-resolution model in a low-resolution part and a detailed part, which is expressed as a wavelet coefficient vector. For embedding, the ratio of the norm of the wavelet coefficient vector to the length of the corresponding edge in the polygonal model (low-resolution part) is modified.

The proposed method of Praun et al. [36] is based on progressive meshes [37] for multiresolution decomposition. The watermark is embedded by adding a scaled basic function value to each vertex coordinate. Extraction is performed by inverting the embedding process using the original and the watermarked mesh. The retrieved watermark is analyzed by calculating a linear correlation. Praun et al. also describe a registration and resampling process necessary for the calculation of the difference vectors between the original and the watermarked mesh: Registration is applied in medical imaging to align two representations of the same object originating from different sources. An overview of different registration methods is given in [38]. For watermarking, registration techniques are important to align the original and the watermarked model properly, which is necessary for the extraction of the embedded information. Praun used the rigid registration method, which considers rigid transformations as proposed by Chen and Medion [39], with uniform scaling as an additional degree of freedom. This algorithm needs a suitable initialization. User interaction is necessary for the initial alignment, especially for cropped images or models with strong symmetries.

Resampling creates a mesh with the topology of the original and the geometry of the watermarked mesh. For each vertex $\mathbf{v}[i]$ on the original mesh, the corresponding point $\hat{\mathbf{v}}[i]$ is calculated. Obviously, registration is necessary before resampling. Praun et al. describe this as an energy minimization problem, where the energy is composed of the distance of the meshes, the deformations of the original mesh by using a spring model, and surface flipping as a penalty.

The whole process is shown in the diagram in Figure 6.9. This scheme is adopted in recent watermarking techniques: Yin et al. [40] use Guskov's relaxation operator for the construction of the multiresolution representation of meshes. The watermark is embedded into the coarser levels of the mesh pyramid. Embedding is related to Praun's method [36], but the vertex coordinates are treated independently. Ohbuchi et al. also presented a watermarking scheme that is based on registration and resampling [41].

Additional approaches to mesh watermarking   Benedens [42, 43] proposes new methods for registration and watermarking of three-dimensional models. Before applying rigid transformation, a preregistration step is introduced. The objective of this preregistration is to eliminate global affine transformations. A principle component analysis (PCA) is applied to reorient the watermarked mesh. Although this is robust against lowpass filtering or polygon reduction, it is sensitive to cropping. Also, while Euclidean transformations and uniform scaling can be handled by this method, this is not the case for general affine and elastic deformations.

Benedens proposed two measures that allow the calculation of affine transformations. Affine transformations maximize (respectively minimize) the following measures: Maximum compactness is defined by the affine transform $\mathbf{A}$ applied to the mesh $\mathcal{M}$ which maximizes $F = \frac{\text{volume}}{\text{area}}$ where volume refers to the volume and area to the surface of the mesh $T_{\mathbf{A}}(M)$. Minimum relative scatter is defined by the affine transform $\mathbf{A}$ applied to the mesh M with vertices $\mathbf{v}[i]$ which minimizes $F = -\frac{\text{variance}}{\text{volume}}$, where variance $= \frac{1}{n}\sum_{i=1}^{n}(\mathbf{A}\mathbf{v}[i] - \overline{\mathbf{v}})^2$.

In the watermarking scheme, which is based on free-form deformations (FFD) [44], strong robustness is achieved. Here watermarks even survive

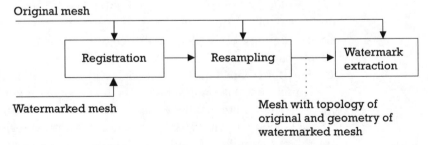

Figure 6.9 The registration and resampling operations proposed by Praun et. al [36]. Registration orients and scales the watermarked mesh to obtain congruent meshes, while resampling uses the original mesh to create the same topology in the registered mesh.

D/A and A/D conversion. D/A conversion experiments were performed by stereo lithography plots. The physical model was reconstructed using software that was developed in the Department of Cognitive Computing and Medical Imaging at the Fraunhofer Institute for Computer Graphics, Darmstadt, Germany [45, 46]. The watermarking method distributes feature points in the original mesh by maximizing their minimum distance to their nearest neighbors. These feature points are modified to embed a 1-bit watermark. The feature points are displaced in normal or inverse-normal direction by applying the FFDs. However, FFDs only approximate the desired changes, and limiting the corresponding vertex movement is somewhat difficult.

CAD data watermarking was proposed by Ohbuchi et al. [47], who suggested a watermarking method based on reparametrization of NURBS curves. The approach suggested by Benedens [29] changes the geometry of the models, making it unsuitable for application areas where changes of the geometry are not allowed.

MPEG 4 uses audio visual objects (AVO). Because of the fact that besides natural data synthetic (computer-generated) data is also considered (by SNHC), watermarking of MPEG 4 objects is also related to watermarking of three-dimensional models. MPEG 4 bit-stream watermarking (watermarking after compression) is discussed by Bartolini et al. [48].

Three-dimensional voxels   Research on three-dimensional voxel watermarking thus far has been limited. Either image watermarking methods are applied directly to the individual slices or the image watermarking algorithms are extended to three dimensions. Tefas et al. [49] proposes a method for gray-scale (color) voxels which imposes intensity (color) constraints to certain voxel neighborhoods.

6.2.5   Quality

For measuring the effects of changes to three-dimensional models, two different aspects have to be considered. Numerical quality affects usability and is mandatory for data, which results in real-world products and depends on the application as well as on the data itself. For example, parts of a car engine require more accuracy than parts of a chassis. Additionally, the accuracy can vary even for one model: Interface parts underly strong limitations. These changes can be measured using distance metrics (e.g., the Euclidean or the Hausdorff distance metric) or by examination of the volumes or surface areas. Considering fairness criteria (e.g., curvature differences), the quality of tessellation, the mesh density, or any degeneracies (such as self-intersections) allows further quality statements. One tool for measuring the numerical quality is the so-called measuring error between surfaces using the Hausdorff distance (MESH) software from Ecole Polytechnique Fédérale de Lansanne (EPFL) [50].

Visual quality is difficult to measure. On the one hand, measuring the quality of changes to three-dimensional models is related to two-dimensional images through the output of rendered images. On the other hand, two-dimensional images are static representations, where lighting and viewing conditions have been fixed. This is typically not the case for three-dimensional models, which can be rendered or viewed under various (and dynamic) conditions. Lighting conditions may change as a result of such parameterization; objects in a scene may *interact* (e.g., intersection), while different textures and shading parameters will have other visual effects. So far, subjective measures are used to evaluate visual quality (see Section 7.3.1). However, different approaches for measuring the objective quality are suggested in [17, 51, 52].

6.3   Digital watermarking for two-dimensional geometry data

Algorithms for two-dimensional geometry data have also been proposed. Typical applications are areas in which vectorized representations play an important role. Geographical information systems (GIS) are only one example of such applications. This area is related to watermarking of three-dimensional data. An approach for watermarking GIS data was proposed by Voigt and Busch [53]. The algorithm is based on the patchwork technique as described in Section 2.3.3. This algorithm is robust against attacks that are in the tolerance range of the underlying data.

6.4   Digital watermarking for formatted text

In contrast to the methods for the data types previously described in this chapter, text has an outstanding property: For media types like images, video, or audio, the watermark is directly embedded in the content of the data as noise. Embedding a watermark directly in the content of text requires a change in the grammatical structure or in the semantic. But the result must be semantically equivalent. A fully automatic solution must be able to understand the grammar and the semantics. Different embedding strategies have been developed for hiding information in text which include manipulations of representations (which permit variations without changing the semantics).

6.4.1   Application requirements

Possible application scenarios are fingerprinting, document authentication, copyright protection, or meta data labeling (e.g., storing archive information). Different operations on textual representations can be identified: Copying is often applied to text for distribution of documents. The traditional photocopier or document image scanner is complemented by the digital copying machine, the combination of scanner and printer. Typesetting is another form of distribution. The text is retyped, for example, for digital archiving or distribution of paper documents. Optical character recognition (OCR) supports the conversion of paper documents (documents in the analog world) to digital files.

6.4.2   Algorithms

Text can be viewed on different abstraction levels. For the purposes of this discussion primarily, texts distributed in printed form are considered. This printed text consists of black dots printed on a white background.

Considering text as an image is straightforward when dealing with scanned documents. Significant numbers of documents are digitally archived by scanning and stored in a rasterized format. Applying digital image watermarking techniques is also straightforward. However, most of the image watermarking techniques have been designed for color or gray-scale images. There are only a few methods published for watermarking binary images like [54] or [55]. Both methods embed information by changing the number of black and white pixels in certain image regions (see Secton 4.3).

The proposed method of Mei et al. [56] was developed for binary raster image representations of text with the aim to reduce visual artifacts. In the first step, the input image is divided into boundary patterns of the characters. One hundred different pairs of boundary patterns are distinguished to reduce visible artifacts. The pairs are visually similar patterns but with different pixel distributions. Second, the boundary of a letter is divided into segments 5 pixels in length. According to the embedded information, the boundary is changed to one of the two patterns of the similar pair.

Considering text as formatted letters or symbols is the first abstraction level. The pixels printed on a piece of paper do not follow a Gaussian distribution. Thus, the choice of the position of the flipping pixel is directly connected with the visual quality of the watermarked image. This is already taken into consideration in [55, 56]. The next abstraction level is the development of watermarking methods that embed information at the representation level. This method was proposed by Brassil et al. [57, 58]. They proposed two different approaches for embedding information: First, changing the spacing of words or lines can be used as an information carrier. Second, changing the format is another possibility to embed information. For example, different fonts or different font sizes can be used.

For the previous abstraction levels of watermarking text documents, re-typing and applying OCR software will remove the watermark. This is not the case for watermarks embedded in plain text.

Considering text as words with a certain meaning is based on using synonyms [59] and on changing the syntactic structure of sentences [60]. This approach requires a semantic analysis and understanding of the given text to choose the proper synonyms and to change the syntactic structures. This watermarking technique also has limitations. If the authors consider their text as art, they will not allow any changes to the text. Also, even a slight modification can have disastrous consequences (e.g., for the parties involved in contracts or other legal documents).

6.5 Music scores

Music scores can be considered similar to text documents. As with text documents, they can be distributed in different formats. Music scores are typically distributed as analog documents (i.e., as printed music scores). Obviously, other formats are also possible, such as Musical Instrument Digital Interface (MIDI) or (typically vendor-specific) music-editing software formats for programs like Finale, Score, or Sibelius.

6.5.1 Application requirements

Concerning the application scenarios, there are also parallels between music scores and text documents. Possible application scenarios for music score watermarking systems are also fingerprinting, document authentication, copyright protection, or meta data labeling. The typical operations applied to music scores are also similar to text: copying, typesetting, and optical music recognition (OMR), which is the musical equivalent to OCR.

At least in the case of German copyright law §53 IV UrhG, the protection afforded music is different from that for other types of content for historic reasons. This results in the protection of editions in addition to the actual opus itself. Therefore, publishers are also interested in protecting individual editions of an opus.

6.5.2 Algorithms

Algorithms for watermarking musical representation are mainly focused on music scores and on MIDI files. Yamaha even announced the product Mid-Stamp in 1998 [61]. Truly robust watermarking must be considered a rather difficult task given the discrete nature of musical symbols. Any changes of the note length, its frequency, and the rest's duration will not be imperceptible. Imperceptible embedding must be based on the redundancy of the MIDI representation. This can also be used by an attacker to remove the embedded watermark. By converting the MIDI file into a standard representation, the embedded information will be erased.

Considering music score watermarking, the level of abstraction determines the information carriers used by watermarking methods. The strong parallels to text watermarking are obvious. These parallels are also visible in the different methods proposed for watermarking music scores.

Considering music scores as images   Funk and Schmucker [62] adapted Zhao's method [54] [see (4.3)] to music scores. They also changed the ratio of the black and white pixels in certain image areas. For improving the quality, they proposed using only horizontal line segments, like staff lines or slurs, for embedding the watermark. This watermarking technique offers high capacity. However, its major drawback is its lack of robustness against transforms.

Considering music scores as represented information guarantees a certain capacity and was proposed by Monsignori et al. [63]. Their approach can be considered as a modulation of the staff line thickness. This approach has increased robustness, and 160 bits can be embedded in a single staff line. This approach is robust against the cropping of individual staff lines. However, an attacker can attempt to change the staff line thickness by applying standard image processing procedures.

Monsignori et al. [64] also suggested another approach, referred to as *staff line masking*. Diagonal lines are hidden in a staff line by introducing gaps in the staff lines. The information is encoded by the angle between the diagonal lines and the staff lines.

A watermarking algorithm that embeds information in musical symbols was proposed by Busch et al. [65]. Features of musical symbols, such as distances between notes or the width of note heads, are used for embedding information. The major drawback of this method is its capacity, which is constrained by the number of symbols available on a music sheet. Embedding one watermark in a single staff line is not typically possible.

The method proposed by Schmucker [66] is also based on the staff lines. But instead of modifying the staff lines' thickness, an image warping technique applies slight geometrical distortion to the staff lines. Again, the capacity of this method is critical. Experimental results showed that only 2 bits can be embedded in each staff line by using this method while retaining adequate visual quality.

Considering music scores as information differs from text watermarking. The apparent redundancy is only of theoretical value: For example, it is possible from a theoretical point of view to split a note into two notes with the same frequency, but only with half the length and an additional symbol which indicates that musicians have to connect these notes; however, musicians will not accept this kind of modification. The reason for this is the fact that the visual representation of music is translated into music while musicians read the representation. Two different representations will be considered as having two different musical semantics by musicians. This is independent from the mathematical redundancy. There is also no syntactical hierarchy that can be used for watermarking because the horizontal position of notes corresponds directly to the temporal relationship between notes.

References

[1] Cheveau, L., E. Goray, and R. Salmon, *Watermarking—Summary Results of the EBU Test*, Technical Report 286, European Broadcasting Union, Grand-Saconnex, Switzerland, March 2001.

[2] Kalker, T., et al. "A Video Watermarking System for Broadcast Monitoring," in P. W. Wong and E. J. Delp, (eds.), *Proceedings of Electronic Imaging '99, Security and Watermarking of Multimedia Content*, Vol. 3657, January 1999, pp. 103–112.

[3] Busch, C., W. Funk, and S. Wolthusen, "Digital Watermarking: From Concepts to Real-Time Video Applications," *IEEE Computer Graphics and Applications*, Vol. 19, No. 1, January 1999, pp. 25–35.

[4] Burgett, S., E. Koch, and J. Zhao, "Copyright Labeling of Digitized Image Data," *IEEE Communications Magazine*, Vol. 36, No. 3 March 1998, pp. 94–100.

[5] Su, K., D. Kundur, and D. Hatzinakos, "A Novel Approach to Collusion-Resistant Video Watermarking," in E. J. Delp and P. W. Wong, (eds.), *Proceedings of Electronic Imaging 2002, Security and Watermarking of Multimedia Contents IV*, San Jose, CA, January 2002, pp. 491–502.

[6] Deguillaume, F., G. Csurka, and T. Pun, "Countermeasures for Unintentional and Intentional Video Watermarking Attacks," in P. W. Wong and E. J. Delp, (eds.), *Proceedings of Electronic Imaging 2000, Security and Watermarking of Multimedia Contents II*, Vol. 3971, San Jose, CA, January 2000.

[7] Niu, X., M. Schmucker, and C. Busch, "Video Watermarking Resistance to Rotation, Scaling and Translation," in E. J. Delp and P. W. Wong, (eds.), *Proceedings of Electronic Imaging 2002, Security and Watermarking of Multimedia Contents IV*, Vol. 4675 of *Proceedings of SPIE*, San Jose, CA, January 2002, pp. 512–519.

[8] Lagendijk, R. L., I. Setyawan, and G. Kakes, "Synchronization-Insensitive Video Watermarking Using Structural Noise Pattern," in E. J. Delp and P. W. Wong, (eds.), *Proceedings of Electronic Imaging 2002, Security and Watermarking of Multimedia Contents IV*, San Jose, CA, January 2002, pp. 520–530.

[9] Jordan, F., M. Kutter, and T. Ebrahimi, *Proposal of a Watermarking Technique to Hide/Retrieve Copyright Data in Video*, Technical Report M2281, ISO/IEC JTC1/SC29/WG11 MPEG-4 meeting, Stockholm, July 1997.

[10] Hartung, F., and B. Girod, "Digital Watermarking of Raw and Compressed Video," *Proceedings of the European Symposium on Advanced Imaging and Network Technologies*, Berlin, October 1996.

[11] Lohman, G., *Volumetric Image Analysis*, New York: Wiley-Teubner, 1998.

[12] Watt, A., *3D Computer Graphics*, Reading, MA: Addision-Wesley, 1995.

[13] Foley, J. D., and A. Van Dam, *Computer Graphics: Principles and Practice*, Reading, MA: Addison-Wesley, 1989.

[14] Encarnacao, J., W. Strasser, and R. Klein, *Graphische Datenverarbeitung: Gerätetechnik, Programmierung und Anwendung graphischer Systeme*, Vol. 1, Munich: 4th ed., R. Oldenbourg Verlag, 1996.

[15] Piegl, L., and W. Tiller, *The NURBS Book*, Heidelberg: Springer-Verlag, 1997.

[16] Mortenson, M. E., *Geometric Modeling*, 2nd ed., New York: John Wiley & Sons, 1997.

[17] Kobbelt, L., "Multiresolution Techniques," in G. Farin, J. Hoschek, and M. S. Kim, (eds.), *Handbook of Computer Aided Geometric Design*, Amsterdam: Elsevier, 2002.

[18] Gotsman, C., S. Gumhold, and L. P. Kobbelt, "Simplification and Compression of 3D Meshes," in A. Iske, E. Quak, and M. Floater, (eds.), *Tutorials on Multiresolution in Geometric Modeling*, Heidelberg: Springer-Verlag, 2002.

[19] Hoppe, H., et al., "Surface Reconstruction from Unorganized Points," *Computer Graphics (SIGGRAPH '92 Proceedings)*, Vol. 26, No. 2, July 1992, pp. 71–78.

[20] Benedens, O., "Geometry-Based Watermarking of 3D Models," *IEEE Computer Graphics and Applications*, Vol. 19, No. 1, January 1999, pp. 46–55. (Special issue on image security.)

[21] Fridrich, J., "Invertible Authentication," in Ed. J. Delp and P. W. Wong, (eds.) *Proceedings of Electronic Imaging 2001, Security and Watermarking of Multimedia Contents III*, San Jose, CA, January 2001, pp. 197–208.

[22] Nikolaidis, N., and I. Pitas, *3-D Image Processing Algorithms*, New York: John Wiley & Sons, 2000.

[23] Ohbuchi, R., H. Masuda, and M. Aono, "Watermarking Multiple Object Types in Three-Dimensional Model," in J. Dittmann, P. Wohlmacher, P. Horster, and R. Steinmetz, (eds.), *Multimedia and Security—Workshop at ACM Multimedia '98*, Bristol, U.K., September 1998, pp. 83–91.

[24] Ohbuchi, R., H. Masuada, and M. Aono, "Watermarking Three-Dimensional Polygonal Models Through Geometric and Topological Modifications," *IEEE Journal on Selected Areas in Communications*, Vol. 16, No. 4, May 1998, pp. 551–560.

[25] Cayre, F., and B. M. Macq, "Three-Dimensional Meshed Objects Watermarking," in A. G. Tescher, ed., *Proceedings of the 47th SPIE Annual Meeting, Applications of Digital Image Processing XXIV*, Vol. 4472, San Diego, CA, July 2001.

[26] Puppo, E., and R. Scopigno, "Simplification, LOD and Multiresolution Principles and Applications," in D. Fellner and L. Szirmay-Kalos, (eds.), *EUROGRAPHICS '97 Tutorial Notes*, Eurographics Association, September 1997.

[27] Benedens, O., "Two High Capacity Methods for Embedding Public Watermarks into 3D Polygonal Models," *Proceedings of Multimedia and Security*, Orlando, FL, November 1999, pp. 95–99.

[28] Gueziec, A., et al., "Converting Sets of Polygons to Manifold Surfaces by Cutting and Stitching," in S. Grisson et al., (eds.), *Conference Abstracts and Applications: SIGGRAPH '98*, Orlando, FL, July 1998, pp. 245–245.

[29] Benedens, O., "Affine Invariant Watermarks for 3D Polygonal and NURBS Based Models," in J. Pieprzyk, E. Okamoto, and J. Seberry, (eds.), *Information Security: Proceedings of the Third International Workshop (ISW 2000)*, Vol. 1975 of *Lecture Notes in Computer Science*, Wollongong, Australia: Springer-Verlag, December 2000, pp. 15–29.

[30] Benedens, O., and C. Busch, "Towards Blind Detection of Robust Watermarks in Polygonal Models," in M. Gross and F. R. A. Hopgood, (eds.), *Eurographics Conference Proceedings*, Interlaken, Switzerland, August 2000, pp. 199–209.

[31] Wagner, M., "Robust Watermarking of Polygonal Meshes," in R. Martin and W. Wang, (eds.), *Proceedings of the Conference on Geometric Modeling and Processing (GMP-00)*, Hong Kong, April 2000, pp. 201–208.

[32] Nielson, G. M., and T. A. Foley, "A Survey of Applications of an Affine Invariant Norm," in T. Lyche and L. L. Schumaker, (eds.), *Mathematical Methods in Computer Aided Geometric Design*, Boston: Academic Press, 1989, pp. 445–467.

[33] Aspert, N., et al., "Steganography for Three-Dimensional Polygonal Meshes," *Proceedings of the 47th SPIE Annual Meeting, Applications of Digital Image Processing XXV*, Seattle, July 2002.

[34] Kanai, S., H. Date, and T. Kishinami, "Digital Watermarking for 3D Polygons Using Multiresolution Wavelet Decomposition," *Proceedings of the Sixth IFIP WG 5.2 International Workshop on Geometric Modeling: Fundamentals and Applications*, Tokyo, Japan: Kluwer Academic Publishers, December 1998, pp. 296–307.

[35] Stollnitz, E. J., T. D. DeRose, and D. H. Salesin, *Wavelets for Computer Graphics: Theory and Applications*, 1st ed., San Francisco, CA: Morgann Kaufmann, 1996.

[36] Praun, E., H. Hoppe, and A. Finkelstein, "Robust Mesh Watermarking," in A. Rockwood, (ed.), *Proceedings of the Conference on Computer Graphics (SIGGRAPH '99)*, Los Angeles, August 1999, pp. 49–56.

[37] Hoppe, H., "Progressive Meshes," in H. Rushmeier, ed., *SIGGRAPH 96 Conference Proceedings*, New Orleans, August 1996, pp. 99–108.

[38] Antoine Maintz, J. B., and M. A. Viergever, "A Survey of Medical Image Registration," *Medical Image Analysis*, Vol. 2, No. 1, April 1998, pp. 1–36.

[39] Chen, Y., and G. Medion, "Object Modelling by Registration of Multiple Range Images," *Image and Vision Computing*, Vol. 10, No. 3, April 1992, pp. 145–155.

[40] Yin, K., et al., "Robust Mesh Watermarking Based on Multiresolution Processing," *Computers and Graphics*, Vol. 25, No. 3, June 2001, pp. 409–420.

[41] Ohbuchi, R., A. Mukaiyama, and S. Takahashi, "A Frequency-Domain Approach to Watermarking 3d Shapes," in G. Drettakis and H.-P. Seidel, (eds.), *EUROGRAPHICS 2002*, Seattle, July 2002.

[42] Benedens, O., "3D Watermarking-Algorithms in Context of OpenSGPlus, *Workshop Multimedia Security and Watermarking*, Darmstadt, Germany: Fraunhofer-IGD, February 2002.

[43] Benedens, O., "Robust Watermarking and Affine Registration of 3D Meshes," *Information Hiding, 5th International Workshop*, in F. A. P. Petitcolas (ed.), *Lecture Notes in Computer Science*, Vol. 2578, Noordwijkerhout, the Netherlands: Springer-Verlag, October 7–9, 2002.

[44] Hsu, W., J. Hughes, and H. Kaufman, "Direct Manipulation of Free-Form Deformations," *SIGGRAPH 92 Proceedings*, Chicago, July 1992, pp. 177–182.

[45] Neugebauer, P. J., and K. Klein, "Adaptive Triangulation of Objects Reconstructed from Multiple Range Images," *IEEE Visualization '97, Late Breaking Hot Topics*, Phoenix, October 1997.

[46] Neugebauer, P. J., "Reconstruction of Real-World Objects via Simultaneous Registration and Robust Combination of Multiple Range Images," *International Journal of Shape Modeling*, Vol. 3, Nos. 1&2, 1997, pp. 71–90.

[47] Ohbuchi, R., H. Masuda, and M. Aono, "A Shape-Preserving Data Embedding Algorithm for NURBS Curves and Surfaces," in R. Werner, (ed.), *Proceedings of the 1999 International Conference on Computer Graphics*, Canmore, Alberta, Canada, June 1999, pp. 180–187.

[48] Bartolini, F., et al., "MPEG-4 Video Data Protection for Internet Distribution," in S. Palazzo, (ed.), *Evolutionary Trends of the Internet, Thyrrhenian International Workshop on Digital Communications (IWDC 2001)*, Vol. 2170 of *Lecture Notes in Computer Science*, Heidelberg: Springer Verlag, September 2001, pp. 713–720.

[49] Tefas, A., G. Louizis, and I. Pitas, "3D Image Watermarking Robust to Geometric Distortions, *Proceedings of the 2002 IEEE International Conference on Acoustics, Speech, and Signal Processing (ICASSP)*, Orlando, FL, May 2002.

[50] Aspert, N., D. Santa-Cruz, and T. Ebrahimi, "MESH: Measuring Error Between Surfaces Using the Hausdorff Distance," *Proceedings of the IEEE International Conference on Multimedia and Expo 2002 (ICME)*, Vol. 1, August 2002, pp. 705–708.

[51] Farin, G., *Curves and Surfaces for CAGD: a Practical Guide*, 5th ed., San Francisco: Morgan Kaufmann Publishers, 2001.

[52] Zhou, L., and A. Pang, "Metrics and Visualization Tools for Surface Mesh Comparison," in R. F. Erbacher et al., (eds.), *Proceedings of Electronic Imaging 2001, Visual Data Exploration and Analysis VIII*, San Jose, CA, January 2001.

[53] Voigt, M., and C. Busch, "Watermarking 2D-Vector Data for Geographical Information Systems," in P. W. Wong and E. J. Delp, (eds.), *Security and Watermarking of Mulitmedia Contents IV (SPIE 2002)*, San Jose, CA, January 2002.

[54] Zhao, J., and E. Koch, "Embedding Robust Labels into Images for Copyright Protection," in K. Brunnstein and P. P. Sint, (eds.), *Intellectual Property Rights and New Technologies. Proceedings of the KnowRight '95 Conference*, Vienna, Austria, 1995, pp. 242–251.

[55] Wu, M., E. Tang, and B. Liu, "Data Hiding in Digital Binary Images," *Proceedings of the 2000 IEEE International Conference on Multimedia and Expo (ICME 2000)*, New York, July 2000, pp. 393–396.

[56] Mei, Q., E. K. Wong, and N. Memon, "Data Hiding in Binary Text Documents," in E. J. Delp and P. W. Wong, (eds.), *Proceedings of Electronic Imaging 2001, Security and Watermarking of Multimedia Contents III*, San Jose, CA, January 2001, pp. 369–375.

[57] Brassil, J., et al., "Electronic Marking and Identification Techniques to Discourage Document Copying," *Thirteenth Annual Joint Conference of the IEEE Computer and Communications Societies (IEEE Infocom '94)*, Toronto, Canada, June 1994. pp. 1278–1287.

[58] Low, S. H., and N. F. Maxemchuk, "Performance Comparison of Two Text Marking Methods," *IEEE Journal on Selected Areas in Communications*, Vol. 16, No. 4, May 1998, pp. 561–572.

[59] Potter, T., "Progress in Steganography: Secret Writing for a Digital Age," *Information Security Bulletin (The International Journal for IT Security Professionals)*, Vol. 4, Issue 7, November 1999, pp. 15–33.

[60] Atallah, M. J., et al., "Natural Language Watermarking: Design, Analysis, and a Proof-of-Concept Implementation," in I. S. Moskowitz, (ed.), *Information Hiding: 4th International Workshop*, Vol. 2137 of *Lecture Notes in Computer Science*, Portland, OR, August 2001, pp. 185–199.

[61] Midstamp: Watertight watermarking.

[62] Funk, W., and M. Schmucker, "High Capacity Information Hiding in Music Scores," in P. Nesi, P. Bellini, and C. Busch, (eds.), *Proceedings of First International Conference on WEB Delivering of Music (Wedelmusic 2001)*, Florence, Italy, November 2001, pp. 12–19.

[63] Monsignori, M., P. Nesi, and M. B. Spinu, "A High Capacity Technique for Watermarking Music Sheet While Printing," in J.-L. Dugelay and K. Rose, (eds.), *Proceedings of the 2001 IEEE Fourth Workshop on Multimedia Signal Processing*, Cannes, France, October 2001, pp. 493–498.

[64] Monsignori, M., P. Nesi, and M. B. Spinu, "Watermarking Music Sheet While Printing," in P. Nesi, P. Bellini, and C. Busch, (eds.), *Proceedings of First International Conference on WEB Delivering of Music (Wedelmusic 2001)*, November 2001, pp. 28–35.

[65] Busch, C., A. Pant, and M. Schmucker, "Digital Watermarking for the Protection of Music Scores," in P. W. Wong and E. J. Delp, (eds.), *Proceedings of Electronic Imaging 2001, Security and Watermarking of Multimedia Contents III*, Vol. 4314, Bellingham, WA, January 2001.

[66] Schmucker, M., "Capacity Improvement for a Blind Symbolic Music Score Watermarking Technique," in E. J. Delp and P. W. Wong, (eds.), *Proceedings of Electronic Imaging 2002, Security and Watermarking of Mulitmedia Contents IV*, January 2002, pp. 206–213.

CHAPTER

7

Contents

Attacks and benchmarks of digital watermarking systems

As discussed in Chapter 3, watermarking systems can be applied in a variety of areas. Depending on the use of the watermarking method, the security properties of the digital watermarks have to fulfill different requirements, some of which were also qualitatively mentioned in Chapter 3. Therefore, according to the design of the underlying application, the identification of possible risks with respect to the usage of a watermarking algorithm is of particular importance. The risk analysis not only reveals security flaws of the application scenario as a whole, but also determines the selection of an appropiate watermarking algorithm.

This chapter starts with an approach to identify security requirements in Section 7.1. The security of the watermarking system can be attacked in several ways. This not only depends on the effect the attack could produce, but on the assumptions that can be made about an attacker in a specific application scenario. Therefore, currently known attacks and a classification thereof are the subject of Section 7.2. Each class of attacks will then be explained in more detail in a corresponding subsection.

As is the case for the security issues of the watermarks and corresponding algorithms, the quality of the watermarked work and the robustness of the watermarks have to be defined for each application. Quality evaluation is of considerable importance, since the usefulness of digital content is mainly defined by the quality that is presumed as required or desirable for the intended application. Methods for evaluating the quality of watermarked objects are detailed in Section 7.3 for different media types. Because of the impossibility of performance at the highest levels of

147

quality and robustness constraints simultaneously due to interdependency of the relevant parameters, the technical assessment of the watermarking technique used must be performed for each application. Benchmarking concepts to evaluate the watermarking algorithm with regard to the intended application scenario, the specific quality, and robustness requirements are the subjects of Section 7.4.

7.1   Threats and risk analysis

In each application requiring a certain level of security, the weakest point or points of the system are where an adversary—assuming knowledge of the overall defensive system—will attempt circumvention of the implemented security mechanisms or elements lacking in such mechanisms. Therefore, a first step in the identification of underlying risks and possible attacks is that the application and the corresponding use cases of the watermarking system have to be defined carefully. This includes the listing of the participating groups and their allowed permissions to perform certain operations.[1] Concerning watermarking the canonical operations are:

- Embedding of watermarks;
- Detection of watermarks;
- Removal of watermarks.

The removal of watermarks is always an impermissible operation in security-related applications. Otherwise, if the watermark does not serve security-related issues, it is irrelevant for an attacker (e.g., in the case of annotation watermarks as described in Section 3.3). Nevertheless, the watermark has to be robust against processing manipulations which can occur in the specific application. The permissible operations applicable to the individual classes of actors can be collected in the so-called *operational table* (see Table 7.1).

Having specified the allowed permissions for each group in turn defines the requisite security properties of the used watermarking system. Given the possible attacks for each group identified by having an entry "No" in their row of the operational table, a new table can be created. The new table is an excerpt from the operational table which lists the possible attacks according

---

1. This approach is taken from [1].

Table 7.1 Operational Table for Playlist Generation

Watermarking Operation	Embed	Detect	Remove
Author society	Yes	Yes	—
Radio station	No	No	No
Public	—	—	—

to the operation performed and the assumptions about the attacker (see Section 7.2.1).

7.2   Attacks

An attack can be described as any processing that circumvents the intended purpose of the watermarking technique for a given application. According to this definition, watermarking attacks include normal processing operations like lossy compression, D/A and A/D conversion, and sample rate conversion, which may happen in an application and unintentionally destroy the watermark. An attack potentially breaks the robustness of the watermark, which in turn is related to the quality of the *attacked* data. Therefore, an attack is successful if it defeats the watermark technique while retaining the quality required according to the specified constraints of the application scenario.

7.2.1   Classification of attacks

In order to easily identify the attacks, a classification of the attacks into several groups helps both the developer of a watermarking algorithm and the user of the watermarking system in identifying the security requirements, as well as judging the usability of the watermarking technology. This may be of vital importance for an entire application scenario, since for some attacks, no reliable countermeasures are presently known. Watermarking techniques can be foiled in several ways which are a direct consequence of the permitted operations of embedding, detection, and removal of the watermarks.

Embedding of the watermark always implies detection of the watermark. Three major categories of effects rendering watermarking useless during detection can be identified:

1. Watermarks cannot be detected. There are two strategies to obtain this result: To remove the watermark or to misalign the embedded watermark and corresponding detector.

2. False watermarks are detected. This can be accomplished by attacks that perform some kind of embedding of false watermarks.

3. Unauthorized detection of watermarks. Algorithms that are not carefully designed can produce false alarms.

The attack table (see Table 7.2) provides a broad overview of all the possible attacks—grouped according to the scope of the two parameters—results of the attack, and assumptions about the attacker.

Different types of attacks are possible depending on the knowledge of attackers, the tools they have at their disposal, and the availability of watermarked versions of the same or different works. Each row in Table 7.2 corresponds to a different assumption about the attacker and represents a variation of one category. The three columns of the table represent the major classes of attacks. Usually, the class of attacks that produces the "no detection" result is further subdivided [2, 3] into two classes according to the way the intended effect is achieved. *Removal attacks* erase the watermark form the watermarked work without using the key for embedding the watermark. *Desynchronization attacks*[2] misalign the watermark detector and the watermark without removing the watermark information.

In general, the power of the attacks is increased by the knowledge and tools the adversaries have at their disposal. For example if the attackers have no prior knowledge about the algorithm but an embedder and detector, they can check the effect of watermark removal with denoising and filtering tools with the detector by embedding their own watermark with the aid of the embedder. The attacks differ in the complexity of the operations involved and the effort the attackers have to expend. Clearly, attacks requiring no prior knowledge constitute the most general form; these are often based on common signal processing operations. Having access to watermarked copies of the same work with different watermarks or different works with the same

Table 7.2 Table of Attacks

Effect Operation	No Detection Remove/Desynchronize	False Detection Embed	Unauthorized Detection Detect
No knowledge	Signal processing/misalignment	Copy attack	—
Algorithm published	Specific designed attacks	Deadlock attack	—
Marked works	Collusion attacks	Copy attack	—
Detector	Oracle attacks	Copy attack	False alarms
Encoder and detector	Custom-tailored oracle attack	Overmarking	False alarms

2. Also called *presentation* [4] or *detection-disabling* [3] attacks.

watermarks offers the possibility to apply different kinds of so-called *collusion attacks*.

Knowledge of the underlying watermarking algorithm should always be assumed in cryptographic systems (Kerckhoffs' principle [5]), and by extension for any security system. If a detector is available to the adversary, other types of attacks are possible. Depending on the values reported by the detector, one can apply the *sensitivity analysis* or *gradient descent attacks*. If both an embedder and detector are available, even more sophisticated attacks like the *custom-tailored oracle attack* [6] can be applied. These attacks all belong to the class of removal attacks described in the following section; the desynchronisation attack class including specific examples are the subject of Section 7.2.3.

An interesting approach to the circumvention of watermarking techniques is the class of embedding attacks yielding a misinterpretation of the detection results (see Section 7.2.4). Therefore, the term *interpretation attacks* is also used to denote this class [4]. Finally, the detection class of attacks is presented in Section 7.2.5.

7.2.2 Removal attacks and manipulations

The removal of watermarks represents the most obvious form of attacking a watermark. The restoring of the original would be the extreme form of this kind of attack. If attackers have no prior knowledge of an algorithm, they can apply distortions they know most watermarks are vulnerable to. The removal of watermarks can also happen unintentionally due to operations during the preprocessing of the data in certain applications.

Signal processing operations   Assuming a differential between the quality of material and capabilities available to a creator and to the eventual user of the material, it can typically be assumed that the watermarked object is processed in some way during the transmission from the watermark embedder to the watermark detector (see Section 2.3). Since processing of the data, particularly in the audio field, is widespread, the IFPI has specified in considerable detail the robustness required of an audio watermarking technique [7].

Besides the requirements of preserving the quality and a bandwidth of 20 bps of the embedded data channel, robustness of the watermark against a wide range of filtering and processing operations is formulated as a necessary feature of the watermarking technology. An even more detailed example of the catalogue of requirements that have to be met by the watermarking method is the audio broadcast monitoring scenario, discussed in Section 3.5, as specified by the EBU (for video broadcasting, see Section 6.1.1).

Additionally, signal processing manipulations can be used in order to remove watermarks. Even users with no special knowledge of signal processing can apply these operations by using common consumer-grade software products for image and video manipulation or audio editing to perform filtering, denoising, and compression (JPEG, MPEG) operations automatically. This is even more critical if the procedure—such as the one detailed in [4] for removing watermarks—is widely distributed.

Robustness against common signal operations such as the addition of noise or localized signal distortions is often achieved by using spread-spectrum signaling techniques (see Section 2.3) in the design of watermarking algorithms. Spreading the watermark energy over a large spectrum minimizes the spectral density, and one can—without additional consideration of features of the signal to be marked—impose a boundary on the quality degradation of the watermarked object. Since a naive attacker has to add enough noise in order to destroy the watermark, this makes such trivial attacks impractical.

Without specific knowledge of the underlying watermarking algorithm, attackers can apply noise removal techniques with the assumption that the added watermark is noiselike. In [8], it was shown that the Wiener filter is the optimal linear-filtering/noise-removal attack for specific watermarking systems. If the added pattern is independent of the cover object, the work and the watermark are drawn from zero-mean normal distributions $N(0, \sigma)$, and linear correlation is used as a detection statistic. Furthermore, Su and Girod [8] showed the countermeasure against this type of attack by shaping the power spectrum of the added watermark according to the power spectrum of the original work:

$$\phi_{ww}(\omega) = \frac{\sigma_w^2}{\sigma_c^2}\phi_{cc}(\omega) \qquad (7.1)$$

In this case, the power spectrum of the watermark signal $\phi_{ww}(\omega)$ is a scaled version of the power spectrum $\phi_{cc}(\omega)$ of the carrier object, where the σ_w^2 and σ_c^2 represent the variances of the distributions from the watermark and the carrier signal. In turn, if the watermark is designed according to (7.1), its perceptual qualities are very close to the original signal, and it is difficult to estimate (or separate) the watermark from the carrier signal. This is also known as the *power-spectrum condition* (PSC). The adaptation required by this condition can be performed explicitly by estimating the power spectrum of the original object or implicitly by embedding the watermark in the compressed domain with automatic adaptation to the power spectrum [9].

Two special signal processing attacks were presented by Craver et al. [10] in the course of the Secure Digital Music Initiative (SDMI) challenge.

They are discussed in the following within the framework set forth by the SDMI challenge.

Watermarking technology is one of the key components of the SDMI system [11] to protect the digital representation of music. In September 2000, SDMI initiated a public challenge to test technologies proposed to be used by SDMI including four watermarking technologies (denoted by the letters A, B, C, and F). For the watermark challenge, SDMI provided three files:[3]

> *File 1:* The original song;
> *File 2:* File 1, watermarked;
> *File 3:* Another watermarked song.

The task of the attacker was to produce a file from file 3, transparent in terms of quality but this is where the watermark detector fails. Furthermore, an oracle was provided by SDMI, where the submissions were judged on success or failure with a binary decision variable.

In this sense, the attack mechanism cannot be considered a pure signal processing attack because of the additional on-line oracle and the access to at least one original and corresponding watermarked file. Nevertheless, the way of removing the watermark from the watermarked songs was done with signal processing methods without knowledge of the underlying watermarking methods. Having access to the original (file 1) offered the attacker the possibility to analyze the difference signal[4] between the watermarked (file 2) and the original signal.

In the attack on challenge B, the difference of the fast fourier transform (FFT) magnitudes of both signals revealed two notches around 2.8 and 3.5 kHz for different segments of the watermarked sample. The attack then filled those notches with random but bounded coefficient values.

In the attack on challenge C, the difference signal revealed narrowband signal centered around 1.35 kHz. Two attacks were performed in challenge C. The first one shifted the pitch of the audio by about a quarter tone (a type of desynchronization attack; see the next section) to move the bursts away from the center frequency. Another one applied a bandstop filter with the center frequency of 1.35 kHz.

Specific designed attacks   Having knowledge of the underlying algorithm, in contrast to the above, enables the attacker to design an attack specific for

---

3. The files were 2 min long in CD format (44.1-kHz sampling rate, 16-bit amplitude resolution).

4. This can be computed in the time or frequency domain.

an algorithm or a class of algorithms by finding and exploiting their weaknesses. One of the easiest form of attacks is applying some kind of filtering (see above). A simple lowpass filter can be applied if it is known from the underlying technology that the watermark energy embeds a significant amount of energy in the high-frequency range.

Publishing the watermarking algorithm by the developer is a general principle derived from cryptography and formulated by Auguste Kerckhoffs in 1883 [5]: The security of a cipher or other security mechanism must rely solely on the secret key, not on the secrecy of the algorithm. This in turn enables experts to examine and validate the techniques or to publish potential security flaws. An example of the result of disregarding the Kerckhoffs principle was demonstrated in the attack on challenge A from Craver et al. [10]. During the analysis of the frequency reponse of the watermarking process, Craver et al. discovered that the underlying algorithm is a complex echo-hiding (see Section 5.4.3) system including multiple time-varying echoes. With basic knowledge of the principles of the algorithm, a patent search revealed more technical detail about the pattern used to implement the multiple time-varying echoes.

Moreover, the search provided the attacker with the probable identity of the company that developed the technology. This again showed the validity of Kerckhoffs' desiderata as the main principles are also applicable in the development of reliable watermarking systems.

Another example of a specific (image) watermarking attack against spread-spectrum methods is nonlinear filtering described by Langelaar et al. [12]. The general idea is to estimate the watermark from the watermarked image c_w. In this special case, an experimentally determined 3×3 median filter is used to produce the filtered image $\hat{c}_w$.

$$c'_w = \text{med}_{3 \times 3}(c_w) \tag{7.2}$$

The difference between the watermarked and watermarked filtered image

$$w' = c_w - c'_w \tag{7.3}$$

is a first approximation of the watermark. Before subtraction, the estimated watermark is filtered a second time with a highpass filter and weighted with an experimentally determined scaling factor to yield the final approximation $\hat{w}$ of the watermark (see Figure 7.1).

Due to the use of estimation in finding the putative watermark signal, this class of attacks is also called *estimation-based attacks* [13]. In estimation-based

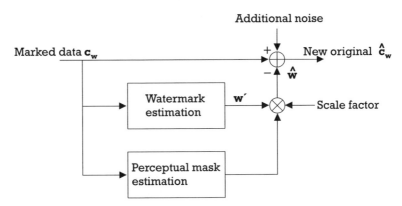

Figure 7.1   Removal by remodulation.

attacks, the knowledge of the watermarking technology and statistics of the original data and the watermark signal are taken into account as seen above.

Having estimated the watermark, an adversary may proceed in different ways, and therefore the attack can be classified as removal, desynchronization, and embedding attacks. A good estimation of the watermark is necessary in two ways. First, the quality of the watermarked object is affected too much if the estimation is not precise enough. Secondly, a rough estimation may not defeat the correlation-based detection of the embedded watermark.

In the context of this section, *remodulation attacks* are a special form of estimation-based attacks trying to remove the watermark by performing a modulation inverse to the embedding of the watermark. The different blocks in the remodulation attack try to cope with the different requirements of quality of the watermarked object and the removal of the watermark. A scale factor is used to adjust between the distortions introduced in the watermarked object. A scale factor $\gamma > 1$ might reduce quality but may lead to a higher certainty of the removal of the watermark. An additional improvement of both goals of the attack can be achieved by calculating the perceptual masking threshold to weight the remodulated watermark. The basic assumption is that the perceptual masking threshold of the watermarked object will be a good approximation of the masking threshold calculated from the original object. This assumption is valid if we assume that the embedding of the watermark to the cover object was parametrized in such a way as to be perceptually transparent.

In the example above, the perceptual mask is substituted with a highpass filter. In order to further decrease the performance of the correlation-based detector, the attacker can add a significant amount of noise in less significant parts of the data (see Figure 7.1). This is an approach that has been demonstrated for image data in [13].

Collusion attacks   Even if attackers have no special knowledge of a specific algorithm or the class it belongs to, they can estimate the watermark or the original if they have more than one watermarked work. In this case, the attackers can apply *collusion attacks*. Estimation of the watermark is possible if they have different works with the same watermark.

In the first case, the attackers have access to $\{\mathbf{c}_{\mathbf{w}i}\}_{i=1}^{n}$ watermarked objects, all watermarked with the same watermark $\mathbf{w}$. They can obtain an approximation of the watermark by averaging the watermarked works[5]:

$$\mathbf{w}' = \frac{1}{n} \sum_{i=1}^{n} \mathbf{c}_{\mathbf{w}i}$$
$$= \mathbf{w} + \frac{1}{n} \sum_{i=1}^{n} \mathbf{c}_{\mathbf{o}i} \tag{7.4}$$

This attack is possible if the added watermark signal is not a function of the original work. Again, a possible countermeasure is to make the watermark dependent on the cover signal. An approximation of the original can be obtained if the attacker has the same work with different watermarks. In fingerprinting applications, where different customer IDs are embedded for identification purposes, building a coalition between different customers can provide access to the same watermarked creation with different watermarks. The same averaging process as described above can be performed which results in estimating the original cover signal:

$$\hat{\mathbf{c}}_{\mathbf{o}} = \frac{1}{n} \sum_{i=1}^{n} \mathbf{c}_{\mathbf{w}i}$$
$$= \mathbf{c}_{\mathbf{o}} + \frac{1}{n} \sum_{i=1}^{n} \mathbf{w}_i \tag{7.5}$$

A method to minimize this problem has been presented by Boneh and Shaw in the form of *collusion-secure codes* [15]. Boneh and Shaw showed that if portions of the coded watermark are identical and carry enough information, at least one of the colluders can be identified. The proof relies on the

5. This has been demonstrated for video applications [14].

assumptions that the identical parts of the coded watermarks will not be affected by the above collusion attack.

A special form of collusion attack to recover the watermark is possible if it is embedded redundantly into distinct segments of the carrier signal. In this case, the carrier signal can be split into different segments, all containing the same watermark. One can regard the different segments of one carrier signal as a set of different carrier signals, all containing the same watermark, and apply the averaging described above.

This type of attack was proposed by Boeuf and Stern [16] for watermarking technology F of the SDMI challenge (see above). Boeuf and Stern presented two ways to remove the mark from file 2 in order to produce the unwatermarked file 3. The basic steps include the estimation of the watermark and a following removal from file 2.

The collusion attack presented by Boeuf and Stern showed that the averaging attack is also possible if the added signal is a function of the original signal as long as the function is known and can be approximated from the watermarked version.

Oracle attacks   Even if the attackers have no knowledge of the algorithm or only one watermarked work, they can apply *oracle attacks* if they have access to a watermark detector [17, 18]. This will be the case in application scenarios where the attackers are allowed to detect watermarks, but not remove them, as in the SDMI scenario, which required the widespread distribution of watermark detectors [11].

Access to the watermark detector can also be achieved by buying products containing the watermark software to embed and detect watermarks such as Adobe Photoshop (using the Digimarc PictureMarc software as a plug-in). The detector can be used as an oracle during the attack on the watermark. Two kinds of oracle attacks relying only on the detector are possible, corresponding with the information returned from the detector.

A binary decision about the presence of the watermark is used in the *sensitivity analysis attack* (see Figure 7.2), whereas the values of the detection statistic are exploited in the *gradient descent attack*. Both attacks use the detector response to find a short path out of the detection region. In the sensitivity analysis attack, it is assumed that this path can be well approximated by the normal to the detection region.

In the gradient descent attack, the direction of the steepest descent is assumed to be a short path out of the detection region. This direction is derived from the gradient of the detection statistic. Both attacks are performed in three steps:

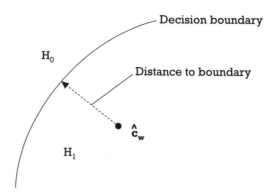

Figure 7.2 Decision boundary for binary hypothesis.

1. Construct an object from the watermarked one, which lies near the detection region boundary. This can be done by applying some kind of signal processing like blurring, filtering, or compression. The constructed object can be a degraded version in comparison to the original. The construction is based on the reponse of the detector. If the output of the detector switches between "yes" and "no" in detecting the watermark, even on slightest modification of the altered object, it lies near the detection boundary.

2. Approximate the path (normal to the detection region or local gradient) out of the detection region. The approximation of the normal to the detection region is done iteratively, where the detection decision is recorded in each iteration. The local gradient is estimated by investigating the change of the detection value if the work is changed smoothly. The approximation of the normal [18] or the search of the local gradient can be implemented as an iterative process.

3. Scale and subtract the direction found in step 2 from the watermarked work (sensitivity analysis attack) or move the work along the direction (gradient descent attack) and repeat step 2.

A special form of an oracle attack called *custom-tailored oracle attack* can be performed if the attackers have access to the embedder and detector [6] (i.e., in an extension to the oracle attacks discussed above). In this case, the attackers embed their own watermarks with the embedder and remove the markings using the oracle attacks described above. The manipulations required to remove one's own watermarks are assumed to be an upper bound for the removal of the watermark to be attacked. This attack is especially useful if the attackers are not allowed to detect the watermark, in contrast to the above scenario.

The sensitivity-analysis attack relies on the assumption that the decision boundary of the binary hypothesis test can be estimated. Performing slight changes until the detector cannot detect the watermark a large number of times any longer yields different points of the decision boundary (see Figure 7.3).

This in turn is used to approximate the normal to boundary in order to find the minimum distortion path out of the detection region. This points to a possible countermeasure, making the decision boundary uncalculable (see Figure 7.4). Tewfik and Mansour present an approach where the decision boundary is modified to have a fractal dimension [19].

To retain the robustness of the watermark, the distance to the new fractal decision boundary is kept constant by modifiying the watermarked object accordingly, which may introduce additional distortions and artifacts. As a result, the modification of the decision boundary has to be adjusted between the two conflicting needs that the decision boundary cannot be approximated even if the attacker has unlimited access to the detector and the requirement of preserving the quality of the watermarked object.

7.2.3   Desynchronization attacks

The aim of the desynchronization[6] as well as the removal attacks is to render the embedded watermark undetectable. Nevertheless, the process of preventing the detection by means of desynchronization attacks is different. Instead

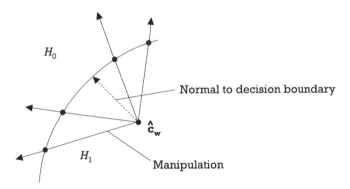

Figure 7.3 Approximation of the decision boundary.

6. Also called *misalignment attacks* [1] .

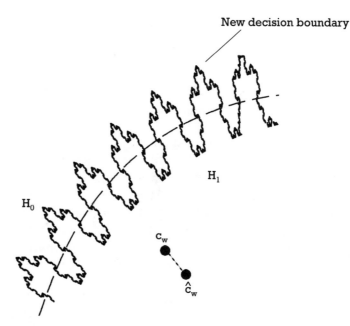

Figure 7.4 Modification of decision boundary.

of erasing the watermark, desynchronization misaligns the embedded watermark and corresponding detector process in such a way in such a way that it is computationally infeasible to perform synchronization prior to detection.

Global and local transformations   Most of the watermarking algorithms, especially those based on correlation, require perfect or near-perfect alignment during detection. Therefore, applying global and local transformations aims at the destruction of the synchronization between the watermark and the detector. Global distortions of watermarked creations include shifting, rotation, and scaling for images and video and delay or time scaling for audio creations. More challenging operations are pitch-preserving time scaling and sample removal in audio, shearing, horizontal reflection, or line removal in images.

While some of these operations can also happen during manipulations, specially developed (benchmark) tools like StirMark (see Section 7.4.2)—in the case of images—perform operations to desynchronize the watermarked creations. Breaking an audio watermarking technology by applying pitch shifting was already presented as a signal processing operation (see the discussion of SDMI attack challenge C in Section 7.2.2).

An example of a time-warping attack was performed in the SDMI attack challenge F by Craver et al. [10]. They warped the time axis by inserting a periodically varying delay defined by the function $f(t)$:

$$t'(t) = t + f(t) \tag{7.6}$$

$$\hat{c}_w(t) = c_w(t'(t)) \tag{7.7}$$

where a represents the original signal and $\tilde{a}$ the attacked one. With the time $t[s]$ and the sampling rate $f_s\left[\frac{\text{samples}}{s}\right]$, the discrete-time function $f(k)$, $k = tf_s$ is defined[7] by (see Figure 7.5):

$$f(k) = \frac{d}{f_s}\left[1 - \cos\left(\frac{p\pi k}{f_s}\right)\right]^2, k = 0, \ldots, l(\mathbf{c_o}) \tag{7.8}$$

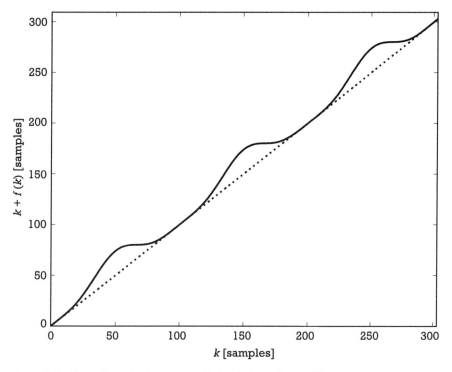

Figure 7.5 Example of the time warp with $f_s = 100$, $p = 2$, $s = 600$.

7. The choice of parameters was made only for illustration purposes.

The number of samples for one period of the function $f(t)$ can be adjusted by p:

$$k = \frac{2}{p} f_s \qquad (7.9)$$

whereas the time delay per period is determined by the maximum number of samples $4d$ delayed:

$$\Delta t = \frac{4d}{f_s} \qquad (7.10)$$

The parameters during the SDMI attack, $f_s = 44.1$ kHz, $d = 6.75$, and $p = 0.602$, were derived from the study of the SDMI challenge A. Therefore, in a period $T = \frac{2}{p} \approx 3.32$ sec the audio files were distorted by 27 samples or $\Delta t \approx 0.6$ ms.

Scrambling attacks   Another kind of desynchronization can be performed by scrambling samples or pieces of the watermarked creation prior to the presentation[8] to a watermark detector. If the watermarked creations are not directly modified but only their presentation, the attacks are performed on a system level[9] that cannot be addressed within the watermarking system itself. Examples include pixel permutations, the Mosaic attack [20], and the use of scrambling and descrambling devices in the case of video.

An advanced form of attack for any watermarking technology independent of the content is not to attack the watermarked version but to create a new work. The assumptions about the attackers are very low, since they do not need to know the algorthim or have access to the detector nor to more than one watermarked work. This kind of generic attack is simulated by the *blind pattern matching* (BPM) *attack* [21]. The rerecording of the creation is done by replacing small pieces of the watermarked creation with perceptually similar pieces from the same watermarked creation or from an external library. A necessary assumption for a successful attack is that the pieces used for substitution contain different watermarks with little correlation to the original watermark. The assumption is valid even if the same watermark is embedded redundantly in a creation. Because this is often done in a block-based manner, the probability that the substitution pieces are carrying the

---

8. These are also referred to as *presentation attacks*.
9. This is the reason they are often called *system level attacks*.

same part of the watermark is low if the size of pieces is one order of magnitude or more smaller than the size of the watermark. The attack was demonstrated by Petitcolas and Kirovski [21] using the same watermarked audio file
and permuting small pieces of it.[10] The basic steps of the algorithm consits of:

Partitioning of the watermarked signal   The watermarked signal $\mathbf{c_w}$ is segmented into $0 \leq i \leq M - 1$ overlapping blocks $\mathbf{c_w}_i$ of equal size N. The
overlap ratio $\eta = 0.25$ between the blocks was choosen to avoid near equivalent perceptual characteristics between adjacent blocks.

Definition of the similarity function   To quantify the similarity between two
blocks $\mathbf{c_w}_i$ and $\mathbf{c_w}_j$, a real positive semidefinite function $\phi(\mathbf{c_w}_i, \mathbf{c_w}_j) \geq 0$ is
defined. In the attack presented in [21], the Euclidian distance between the
blocks was used:

$$\phi(\mathbf{c_w}_i, \mathbf{c_w}_j)^2 = \sum_{k=0}^{N} (\mathbf{c_w}_i[k] - \mathbf{c_w}_j[k])^2 \tag{7.11}$$

Of course, attackers can use other metrics possibly more suited to perceptual quality in calculating perceptual distance.

Pattern matching   Perceptual similarity is identified in this step by the construction of a similarity bit matrix $S = \{S_{ij}\}_{i,j=1}^{M}$ defined by

$$S_{ij} = \begin{cases} 1 & N\alpha^2 \leq \phi(\mathbf{c_w}_i, \mathbf{c_w}_j)^2 \leq N\beta^2 \\ 0 & \text{otherwise} \end{cases} \tag{7.12}$$

The elements $S_{ij} = 1$ identify similar blocks, where all other $S_{ij} = 0$
are classified as dissimilar.[11] The parameters α and β define the acceptable
range of similarity. α defines the maximum similarity, whereas β defines the
minimum similarity. If α is to low, the detection of the watermark is not
affected because of a too strong similarity. If β is set to high, the distortion
introduced by substituting a block with too low similarity affects the quality
of the attacked data too much. Since the BPM attack is performed in the
logarithmic frequency domain, β [dB] can be interpreted as the maximum
noise allowed to be introduced by the substitution of the blocks, and α [dB]
the minimum distance between similar blocks.

---

10. This explains the classification of BPM as a scrambling attack.
11. This includes the equal blocks $S_{ii} = 0$, $i = 1, \ldots, M$.

Block substitution   Having identified all similar blocks according to the pattern matching step, the attacked signal $\hat{c}_w$ is created by pseudorandomly permuting these blocks.

The choise of the block size is a degree of freedom depending on the data presented. Smaller blocks have a higher correlation between original and substitute and therefore do not affect the watermark detection to the same extent. In contrast it may be difficult to find a sufficient number of larger blocks in the required similarity range according to (7.12).

Performing the attack experimentally showed that about half the blocks could be substituted in a 30-sec audio clip within noise margins $\beta = 4.5$ dB and $\alpha = 1.5$ dB. Using the same parameter setup showed a decrease of the normalized correlation to 50% of the expected value. A possible countermeasure against this attack is to identify blocks with no similar counterparts and use only these blocks for embedding.

7.2.4 Embedding attacks

Embedding or ambiguity attacks simulate an embedded watermark even if it is not embedded. The effect of this attack is the false detection of watermarks in contrast to no detection of the removal or desynchronization attacks. Three main variations of this attack will be considered in the following paragraphs according to the assumption that can be made about the attackers.

Copy attack   The aim of the *copy attack* is to copy a watermark from one carrier signal to another. This attack is basically performed in two steps. In the first step an estimation of the watermark from the marked carrier signal is calculated. In the second step the estimated watermark signal is copied from the marked signal to the target carrier signal data to obtain a watermarked version (see Figure 7.6).

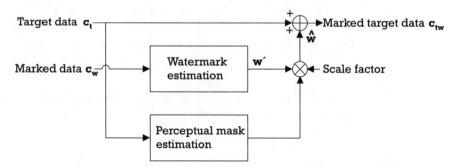

Figure 7.6 The copy attack.

The estimated watermark can be obtained in different ways depending on the assumptions made about the attackers. If the attackers have no prior knowledge of the algorithm but have access to the same object carrying different watermarks, they can perform the collusion attack (see Section 7.2.2). This approximates the original object. The watermark they want to estimate is then obtained by subtracting the estimated original from the corresponding watermarked version [see (7.3)].

The original presentation of the copy attack by Kutter et al. [22] performed a watermark removal attack to obtain the original via spatial domain filtering. A previous approximation of the original is not necessary if the attackers can estimate the watermark directly. This is possible if they use the first version of the collusion attack, where access to different objects with the same watermark is required (see Section 7.2.2).

A possible countermeasure to prevent the copy attack is to establish a link between the watermark and the carrier signal via cryptographic hash functions [1]. This link can be verified during the detection of the watermark. Another possibility may be to make the watermark a function of the original carrier signal. In this case, copying will be more problematic in terms of the quality of the marked target carrier signal.

Overmarking   Overmarking is an operation where a second watermark is embedded in an already marked carrier signal. Both watermarks can be detected independently if, for example, the location where the information will be hidden is determined by the secret key.

This operation can always be performed if the attacker has access to the embedder and detector of the watermarking system. If the intention of the watermark is copyright protection, both parties (the copyright owner and the attacker) may claim ownership. The problem of ownership would be solved in this case if the order of watermark insertion can be proven reliably.

The only advantage the copyright owner has compared to the attacker is access to the true original. Since the attacker has only access to the already watermarked object, the sequence of embedding would be determined by the fact that both parties have to read the watermark from their corresponding putative original. A problem therefore arises if both parties can read their watermarks from the original of the opponent. In this case, a stand-off is created, where the copyright owner has no real advantage over the adversary. This is aim of the so-called *deadlock* or *IBM attack*.[12]

---

12. Occasionally also called *Craver attack*.

Deadlock attack   Different forms of the deadlock attack are possible depending on the possibility of access of the attacker to the original creation. To distinguish between watermarks of the copyright owner and the attacker, the letters c and a, respectively, will be used. The basic assumption of this attack is that the correlation between true and the fraudulent watermarks is very low (this is very likely).

$$C_\tau(\mathbf{w_c}, \mathbf{w_a}) \approx 0 \qquad\qquad (7.13)$$

Furthermore, the watermarked creation $\mathbf{c_w}$ and the fraudulent original $\mathbf{c_a}$ are created according to the following equations:

$$\mathbf{c_w} = E_{K^c}(\mathbf{c_o}, \mathbf{w_c}) = \mathbf{c_o} + \mathbf{w_c} \qquad\qquad (7.14)$$

$$\mathbf{c_a} = E_{K^a}^{-1}(\mathbf{c_w}, \mathbf{w_a}) = \mathbf{c_w} - \mathbf{w_a} \qquad\qquad (7.15)$$

$$\Rightarrow \mathbf{c_w} = E_{K^a}(\mathbf{c_a}, \mathbf{w_a}) \qquad\qquad (7.16)$$

$\mathbf{c_w}$ is constructed via an ordinary embedding process whereas the creation of $\mathbf{c_a}$ is based on the inversion of the embedding process of a watermarking system. It looks like a regular embedding function, and embeds the fraudulent watermark $\mathbf{w_a}$ into the fraudulent original $\mathbf{c_a}$ to yield $\mathbf{c_w}$. In the case of informed detection, Alice can demonstrate that $\mathbf{w_c}$ is embedded in $\mathbf{c_w}$ and the fraudulent original $\mathbf{c_a}$ by building the difference that should be close to $\mathbf{w_c}$ for a robust method:

$$C_\tau(\mathbf{c_w} - \mathbf{c_o}, \mathbf{w_c}) = 1 \qquad\qquad (7.17)$$

$$C_\tau(\mathbf{c_a} - \mathbf{c_o}, \mathbf{w_c}) = C_\tau(\mathbf{w_c}, \mathbf{w_c}) - \underbrace{C_\tau(\mathbf{w_a}, \mathbf{w_c})}_{=0} = 1 \qquad\qquad (7.18)$$

Nevertheless, Bob can also prove that his fraudulent watermark $\mathbf{w_a}$ is embedded in $\mathbf{c_w}$ and the original $\mathbf{c_o}$:

$$C_\tau(\mathbf{c_w} - \mathbf{c_a}, \mathbf{w_a}) = 1 \qquad\qquad (7.19)$$

$$C_\tau(\mathbf{c_o} - \mathbf{c_a}, \mathbf{w_a}) = -\underbrace{C_\tau(\mathbf{w_c}, \mathbf{w_a})}_{=0} + C_\tau(\mathbf{w_a}, \mathbf{w_a}) = 1 \qquad\qquad (7.20)$$

Craver et al. [23] demonstrated this attack in an informed detection system by using the Cox algorithm [24], where the strength of the detection measured by C_τ was nearly identical.

If blind detection is used, the additional requirement that the correlation between the original and assumed watermark is zero [when compared to threshold τ according to (2.5)] has to be made:

$$C_\tau(\mathbf{c_o}, \mathbf{w_c}) = C_\tau(\mathbf{c_a}, \mathbf{w_a}) = 0 \qquad (7.21)$$

Disregarding this assumption would accept the existence of false alarms due to the correlation of the original creation and pseudorandomly generated watermarks. During blind detection, Alice has to demonstrate that her watermark is embedded in $\mathbf{c_w}$ and the fraudulent original $\mathbf{c_a}$:

$$C_\tau(\mathbf{c_w}, \mathbf{w_c}) = \underbrace{C_\tau(\mathbf{c_o}, \mathbf{w_c})}_{=0} + C_\tau(\mathbf{w_c}, \mathbf{w_c}) = 1 \qquad (7.22)$$

$$C_\tau(\mathbf{c_a}, \mathbf{w_c}) = C_\tau(\mathbf{c_w}, \mathbf{w_c}) - \underbrace{C_\tau(\mathbf{w_a}, \mathbf{w_c})}_{=0} = 1 \qquad (7.23)$$

Again, Bob can also prove that his watermark is embedded in $\mathbf{c_w}$ and the original $\mathbf{c_o}$:

$$C_\tau(\mathbf{c_w}, \mathbf{w_a}) = \underbrace{C_\tau(\mathbf{c_a}, \mathbf{w_a})}_{=0} + C_\tau(\mathbf{w_a}, \mathbf{w_a}) = 1 \qquad (7.24)$$

$$C_\tau(\mathbf{c_o}, \mathbf{w_a}) = C_\tau(\mathbf{c_w}, \mathbf{w_a}) - \underbrace{C_\tau(\mathbf{w_c}, \mathbf{w_a})}_{=0} = 1 \qquad (7.25)$$

Therefore, in both detection cases the copyright owner would have no advantage compared to the attacker. This is even more suprising, since the attacker has no access to the true original creation $\mathbf{c_o}$. The vulnerability exploited in producing this ambiguity is the invertibility of the watermarking algorithm according to (7.15). One defensive approach is to make the watermark a function of the original, such that the fraudulent watermark cannot be created without access to the original.

$$\mathbf{c_a} \neq E_{K^a}^{-1}[\mathbf{c_w}, \mathbf{w_a}(\mathbf{c_a})] = \mathbf{c_w} - \mathbf{w_a}(\mathbf{c_a}) \qquad (7.26)$$

According to (7.26), construction of the fraudulent original is not possible because the creation depends on the fraudulent original itself. One example of the function $\mathbf{w_c} = f(\mathbf{c_o})$ is to use the hash $H(\mathbf{c_o})$ over the original as the seed of the pseudonoise (PN) generator.

Another possibility is to use the hash $\mathbf{w_c} = H(\mathbf{c_o})$ as the watermark to be embedded. This is possible for every watermarking algorithm, since the problem of preventing the inversion of the embedding function $E_K(\mathbf{c_o}, \mathbf{w})$ is shifted to the problem of inverting the function $f(\mathbf{c_o})$, which is assumed computationally infeasible when using a cryptographic hash function.

7.2.5   Detection attacks

The opposite of unauthorized embedding is unauthorized detection. Usually, unauthorized detection of the embedded watermark is used as the step before the corresponding removal. On the other hand, *detection attacks* simulate the detection of watermarks even if these watermarks were not inserted before. In this case, the attack produces *false alarms*.

False alarm attacks   Usually, the effect of producing false alarms requires that the attackers have access to the detector. The question to ask is which kind of false alarm should be generated? With request to the copyright protection application, the following question has to be answered. Is the watermark $\mathbf{w}$ embedded in the dataset $\mathbf{c_o}$? In this case, applying a false alarm attack would require the attackers to satisfy the following equation:

$$D_K(\mathbf{c_o}, \mathbf{w}) = \mathbf{w}' \quad \text{and} \quad C_\tau(\mathbf{w}', \mathbf{w}) \geq \tau \tag{7.27}$$

If the watermark is fixed, the attackers can vary the parameters $K, \mathbf{c_o}$. Varying the key K is equivalent to a search of a key for a fixed object $\mathbf{c_o}$. Using a fixed key would require the attackers to change the object until (7.27) is satisfied. With the detector in their hands, this can be acomplished by performing the sensitivity analysis attack described in Section 7.2.2. Of course, both types of approaches should in general be computationally infeasible for the watermarking algorithm.

Considering the number of attacks presented in the last section, it seems to be a rather difficult task for the designer of a watermarking algorithm to cope with all or even a subset of possible attacks. Nevertheless, the success of an attack as well as the applicability of the whole watermarking technology is related to the quality of the attacked or watermarked data. Methods for evaluating the quality are described in detail in the next section.

7.3   Quality evaluation of watermarked objects

Unfortunately, the output quality of a watermarking codec cannot be quantified easily in objective terms. This is especially a problem for evaluating the watermarking methods operating on new data types like three-dimensional models or images of music scores, since quality criteria and corresponding evaluation procedures are not currently known and are therefore the subject of current research efforts. On the other hand, this problem is very similar to the task of evaluating the perceptual quality in the audio, image, and video fields. The research in these fields was driven by the development of highly effective data compression software that preserves the quality at the same time. Distortions introduced during perceptual coding are due to quantization noise added in the coding process. The quantization noise is hidden below the computed perceptual threshold. In watermarking systems, an additional signal carrying information is added likewise. This signal is shaped according to the masking threshold to ensure the quality of the watermarked signal. Conversely, the problems related to the quality of the watermarked carrier signal are expected to be very similar to those in the perceptual coding case. Therefore, an evident approach is to use principles and test procedures already investigated and applied during the development of the algorithms. In general, two types of tests are applied: human subjective assessment techniques and objective evaluation methods.

In this context, subjective tests are still used as a tool for codec quality evaluation. Standardized test procedures have been developed to maximize the reliability of the results of subjective testing. The next chapter describes general methods developed to perform subjective quality testing which can be applied regardless of the media type. Subsequently, subjective and—if available—corresponding objective quality measurement techniques are detailed for perceptual audio, image, and video codecs. This includes approaches to adapt these techniques to the watermarking problem.

7.3.1 Psychophysical methods

Subjective evaluation can be performed in two different ways: testing the transparency of the watermarked items or, in a more general way, rating the quality of the processed items with respect to the reference signal. In the following, an item is called *transparent* if no differences between the original and the watermarked version are perceivable. Otherwise it is called *nontransparent*. All tests involving human beings have in common that they require:

- A specification of the evaluation environment;
- A careful selection of the test material;
- A training phase for the assessors of the test;
- A test phase that consists of a comparison of the coded material against a reference (original);
- A statistical analysis for a quantative interpretation of the results.

The field of *psychophysics*[13] is the science to derive correlations between quantitative variables and qualitative experience of human beings. Corresponding psychophysical methods are used wherever studies involving the judgment of subjects are necessary. Therefore, subjective evaluation methods for watermarked objects can take advantage of these procedures in order to gain quantitative insights into the quality of the watermarked objects.

The two-alternative-forced-choice test   If the impairments introduced by the coding procedure are very small, one can assume transparency of the coded signal. To further validate this hypothesis, a subjective evaluation test for nontransparency can be performed by a so-called *two-alternative-forced-choice test*. In this case, the hypothesis of nontransparency is tested (in contrast to the additional rating described below). A training phase precedes the actual test phase. During the training, test persons compare the original and the watermarked item.

For the actual test, a number of pairs are randomly chosen from the set of possible combinations $\{(c_o, c_o), (c_o, c_w), (c_w, c_o), (c_w, c_w)\}$. For each of these pairs, the subjects are asked whether both items were equal or not. A correct decision about items being equal or different is called a *hit*, so subjects produces a result of the form "k hits of a number of trials." During the test phase, no limit is imposed on the number of repetitions to compare each of the individual items in the pair. Since the evaluation is performed for a group of persons, the hits within this group are summed and taken as the test variable.

A test for nontransparency is performed by trying to reject the transparency hypothesis. Therefore, the following null and corresponding alternative hypotheses are formulated:

H_0: Distortions are not perceivable.
H_1: A subject can perceive distortions in a watermarked item.

What is tested is the ability to detect differences between the original and the watermarked object. Therefore, the test variable is the number of hits k out of the number of pairs n. Under the hypothesis of H_0, the probability to get k hits out of n with detection probability $p = 0.5$—because the subject is simply guessing—is

13. A term coined by the German physicist and psychologist Gustav Fechner (1801–1887).

$$P(k, n, p) = \binom{n}{k} p^k (1 - p)^{n-k} = \binom{n}{k} 0.5^n \qquad (7.28)$$

Therefore, the distribution function of k is a binomial distribution $B(n, p)$. The critical region can be determined according to the following equation by choosing a level of significance α:

$$P(T \in \mathbf{B} \mid H_0) \leq \alpha \qquad (7.29)$$

Choosing the level of significance $\alpha = 0.05$ and applying (7.29) for $n = 40$ pairs lead to a critical region of $\mathbf{B} = \{26, \ldots, 40\}$; that is, if a subject has more then 25 hits, the error probability of *wrongly rejecting* the null hypothesis is 5%.

To ensure independent experiments, the best approach would be to use as many subjects as possible, each one testing only one pair. If the subjects are able to distinguish between the original and the watermarked version, they will do it with a certain probability of detection $p > 0.5$. In general, there exists no knowledge about this detection probability. The parameter $p = 0.5$ corresponds to the null hypothesis H_0, whereas all values of $0.5 < p \leq 1$ correspond to the alternative hypothesis H_1 of nontransparency.

What can be calculated is the so-called operation characteristic (OC) function $\beta(p)$. $\beta(p)$ is a function of the detection probability p and reports the error of *wrongly accepting* H_0 if the alternative hypothesis H_1 is correct:

$$\beta(p) = \begin{cases} 0.95, & p = 0.5 \\ P_p(T \notin \mathbf{B} \mid H_1), & 0.5 < p \leq 1 \end{cases} \qquad (7.30)$$

The adjustment of the OC function $\beta(p) = P_p(T \notin \mathbf{B} \mid H_1)$ is done according to (7.30). By using the OC function, we have the full knowledge of the errors and therefore the quality of the hypothesis test. The quality of the test strongly depends on the number of experiments performed. For example, if the subjects are able to distinguish with a detection probability of $p = 0.7$, the error probability is

$$\beta(p) \approx \begin{cases} 0.617, N = 10 \\ 0.007, N = 100 \end{cases} \qquad (7.31)$$

The quality of the test increases with a higher number of tested pairs (see Figure 7.7). This can be used in order to design the test by calculating the

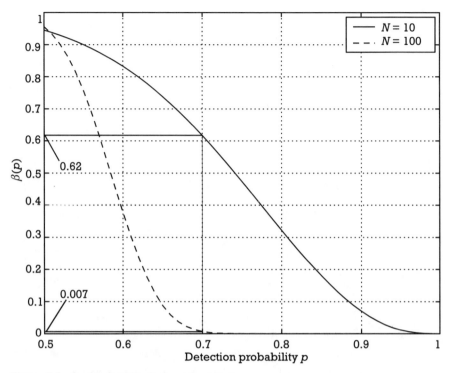

Figure 7.7 Quality function for hypothesis test.

number of pairs needed to ensure the two kind of errors incorporated in
hypothesis testing if the detection probability of the subjects is known.

7.3.2 Audio data

The ITU-R BS.1116 standard   The standard for subjective evaluations of small
impairments of high-quality perceptual audio codecs is specified in the International Telecommunication Union (ITU-R) Recommenodation BS.1116
[25].[14] This procedure is the so-called *double-blind A-B-C triple-stimulus* hidden
reference comparison test. Stimulus A contains the reference signal, whereas
B and C are pseudorandomly selected from the coded and reference signals.
Therefore, either B or C is the hidden reference stimulus. After listening to all
three items, the subject has to decide between B and C as the hidden reference. The remaining signal is the suspected impaired stimulus. This one has to
be graded relative to the reference signal by using the five-grade impairment
scale according to ITU-R BS.562 [26].

14. Published in 1994 and updated in 1997.

Table 7.3 contains absolute and difference grades. The "Grade" column can be treated as a continuous 41-point absolute category rating (ACR) impairment scale. It is used by the listener to grade the impaired signal relative to the reference signal. The stimulus that is identified by the subject as the hidden reference will be assigned the default grade of 5.0. The results of the listening tests are based on the so-called subjective difference grade (SDG) shown in the right column of Table 7.3. It is calculated from the results of this rating by subtracting the score assigned to the actual hidden reference signal from the score assigned to the actual coded signal:

$$SDG = Score_{Signal\ Under\ Test} - Score_{Reference\ Signal} \qquad (7.32)$$

Transparency is assumed if the SDG value is 0, whereas a value of -4.0 is subjectively considered very annoying.

Besides the fact that rigorous subjective listening procedures as described above are still the ultimate quality judgment, they do have some disadvantages:

- The test results are influenced by the variability of the expert listeners. Experiments have shown that the various experts are sensitive to different artifacts [27–29].

- Playback level (SPL) and background noise can introduce undesired masking effects.

- The method of presenting the test items can have a strong influence on the quality (influence of loudspeakers and listening room of the specific site).

- Listening tests are time consuming.

- The equipment necessary to perform listening tests is cost intensive.

Therefore, the need of an automatic perceptual measurement of compressed high-fidelity audio quality has motivated research into development

Table 7.3 ITU-R Five-Grade Impairment Scale

Impairment	Grade	SDG
Imperceptible	5.0	0.0
Perceptible, but not annoying	4.0	−1.0
Slightly annoying	3.0	−2.0
Annoying	2.0	−3.0
Very annoying	1.0	−4.0

of corresponding schemes. In the same way, objective measurement tools are superior to subjective listening tests during the development phase of new audio watermarking algorithms, because of the effort and time that has to be invested.

Objective measurement of high-quality audio   The ultimate goal of objective measurement algorithms is to substitute the subjective listening tests by modeling the listening behavior of human beings. The output of the algorithms for objective measurements is a quality measure consisting of a single number to describe the audibility of the introduced distortions like the SDG in subjective listening tests. The various algorithms for objective measurement of audio quality fit into the general architecture according to Figure 7.8.

A difference measurement technique is used to compare the reference (original) signal and the test (processed, that is, compressed or watermarked) signal. Both the reference signal and the signal under test are processed by an auditory model, which calculates an estimate for the audible signal components. These components can be regarded as the representation of the signals in the human auditory system. The *internal representation* is often related to the masked threshold, which in turn is based on psychoacoustic experiments performed by Zwicker and Fastl [30]. From these two different internal representations of the original and test signals, the *audible difference* is computed. Because the results of the listening tests are judged with a single SDG value, the corresponding measure has to be derived from the audible difference. This is accomplished with the *cognitive model*, which models the processing of the signals by the human brain during the listening tests. The output of the whole system is the *so-called objective difference grade* (ODG), which can be compared to the SDG in the listening test.

This has ultimately led to the adoption of an international standard for the measure of the perceived audio quality (PEAQ), ITU-R BS.1387 [31]. The intention of this standard is to replace the described ITU-R BS.1116

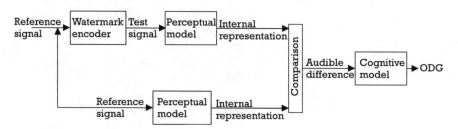

Figure 7.8 General architecture for objective quality measurement of audio data.

standard, which is very sensitive and enables the detection of even small distortions. Since the intention of both the subjective and objective audio quality measurements is to compare the processed audio material with the original signal, the test procedures are only useful in testing high-quality audio. Applying these methods to the evaluation of audio material with lower quality would lead to test results that were mainly directed toward the bottom of the five-grade impairment scale (see Table 7.3) and therefore useless. For this reason, the scope of the ITU-R BS.1116 and ITU-R BS.1387 standards is limited to data rates above 64 Kbps.

Testing watermarked items with reduced quality   In certain watermarking applications, it might be reasonable to use data rates below the 64-Kbps limit. In this case, a problem arises with the evaluation of the quality by both subjective listening tests and objective measurement systems. As noted above, the BS.1116 standard and its objective counterpart BS.1387 are not intended for bit rates below 64 Kbps. This limit might decrease in the future due to advances in the development of high-quality perceptual audio compression codecs. Furthermore, new advanced subjective listening tests termed *multi-stimulus with hidden reference anchors* (MUSHRA) are proposed by an EBU project group [32]. As of late 2002, this test method is in the standardization process of the ITU-R [33]. In contrast to BS.1116, MUSHRA is a *double-blind multistimulus* with hidden reference and hidden anchors. Since the subject will normally easily detect the larger distortions, the usage of a hidden reference makes no sense in this test. The anchors are chosen according to the type of distortions the systems[15] under test are typically introducing.

The difficulty in the evaluation of the impairments is the rating of the relative annoyances of the various artifacts. From this point of view, the subjects have to decide if they prefer one type of impairment over the other. As a result, comparison is not only made between the reference and system under test, but also with all other systems contributing to the test. A grading is performed between the different systems. It is derived by comparing that system to the reference signal as well as to the other signals in each trial. A trial in turn consists of the presentation of the reference signal, the anchors, and all versions of the test signal processed by the systems under test. In contrast to BS.1116, MUSHRA uses an absolute measure of the audio quality directly compared to the reference as well as the anchors. The grading scale in MUSHRA is the five-interval continuous quality scale (CQS), which is divided

---

15. In the context of watermarking, the term *system* is synonymous with different watermarking systems.

into five intervals as shown in Table 7.4. This absolute scale is necessary in order to be able to compare the results with similar tests.

To summarize the different test methods, one has to consider the application and corresponding quality requirements, which should be specified in terms of bit rates. For example, if the audio data are compressed during the application with a specific bit rate, this figure can be used in order to decide which test method is appropriate for evaluating the quality of the watermarked items according to Figure 7.9.

7.3.3 Image data

In this section, the issue of image quality for benchmarking is elaborated. This necessitates a basic definition of quality for the specific context of watermark benchmarking. Keelan [34] defines image quality:

> The quality of an image is defined to be an impression of its merit or excellence, as perceived by an observer neither associated with the act of photography, nor closely involved with the subject matter depicted.

Table 7.4 Five-Interval Continuous Quality Scale (CQS)

Quality	Grade	Internal Numerical Representation
Excellent	5.0	100
Good	4.0	80
Fair	3.0	60
Poor	2.0	40
Bad	1.0	20

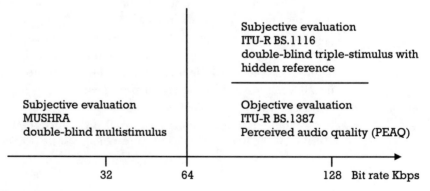

Figure 7.9 Quality measurement methods as a function of the bit rate.

One possible approach to measurement of image quality is to measure the distance between the original and the processed images. With a mathematical norm such as the Euclidean norm ($\|x\|_2 = \sqrt{\sum_i |x_i|^2}$), this is straightforward.

A normalized difference measure, the root-mean-square error (RMSE) can be derived from the Euclidean norm. Simplicity is its major advantage. However, this approach suffers from some drawbacks, the primary defect being that values are absolute and depend on the range of the pixel values. Related measures (e.g., the mean-square error (MSE)) are discussed in [35].

Relative measures do not suffer this obvious drawback: Signal-to-noise ratios (SNR) are used in communication theory to express the relation of the power of the communication signal to the power of the noise signal. Some of these are described in [36]. While these ratios are adequate measures in communication theory, they have an important drawback: These measures are badly correlated to the *perceived* quality.

Different measures have been proposed that model the HVS. When talking about human visual perception, the just noticeable difference (JND) is a measure of the perceptual continuum which is discussed in detail in [34].

Probably the most well known HVS model is Watson's DCT-based visual model [37]. This model considers the varying sensitivity of the HVS to different spatial frequencies and orientations, as well as masking effects like luminance and contrast masking [38].

The human eye is sensitive to luminance differences. However, Weber's law states that the ratio of the just discriminable differences to their stimuli is approximately constant. In addition, the human contrast perception varies with the local image content as well as to spatial frequencies. This concept is described by the contrast sensitivity function (CSF). Other parameters for the CSF are the temporal frequencies, the orientation, viewing distances, and color directions. Nadenau et al. [39] discuss the characteristics of human visual models.

Besides the Watson model, the Sarnoff model as proposed by Lubin [40] and the visual difference predictor by Daly [41] are common models for the human visual system. However, as stated in [42], few comparative studies exist.

7.3.4   Video data

As already mentioned in Section 7.3.3, the CSF also depends on temporal frequencies. In his Ph.D. thesis [43], Winkler discussed objective video quality measures, which he called a perceptual distortion metric (PDM). Winkler's PDM and other objective video quality models were evaluated by the Video Quality Experts Group (VQEG) [44]. One of these models was the peak-signal-to-noise ratio (PSNR), which was not originally included but is commonly accepted as an objective reference measure. This system evaluated the quality of video sequences with different characteristics (e.g., format, temporal information, spatial information, color information). In addition, human viewers performed subjective evaluations. The results were analyzed using statistical methods such as analysis of variance (ANOVA).

One salient point in this evaluation of objective video quality models is the fact that VQEG was not able to propose one specific objective model, because the performance of these models was statistically equivalent to the results of PSNR. In the final report [44], conclusions were summarized as follows:

> No objective measurement system in the test is able to replace subjective testing.

> No objective model outperforms the others in all cases.

> The analysis did not indicate that a method could be proposed for ITU Recommendation.

7.4 Benchmarking

In the context of standardization activities, objective performance metrics are needed to evaluate whether one of the established or emerging watermarking techniques is superior to the available alternative methods. Watermarking algorithms are mainly judged by two evaluation criteria: their ability to preserve the quality of the original carrier signal and the robustness of the embedded watermarks.

Quality is also related to the usefulness of the watermarked data for the application. If the watermarked signals are not satisfactory in terms of quality for the intended application, further considerations about the robustness of the embedded watermarks are superfluous. Methods for quality evaluation are described in Section 7.3 and in Section 6.2.5.

The concept of robustness is obviously and strongly connected to the quality of the watermarked items:

DEFINITION

Watermarks are robust if they cannot be destroyed without affecting the quality of the watermarked object in such a way that it is useless for the application.

As already discussed in Chapter 2, both criteria cannot be maximized simultaneously. It is impossible to ensure the highest quality of the watermarked signal and maximum robustness of the embedded watermarks at the same time. Therefore, as discussed in Chapter 3, evaluation of the usefulness of specific watermarking methods is always to be performed for a specific application and its requirements.

As described above, a number of attacks exist that can destroy the watermark itself or the information carried by the watermarking application. However, further evaluation criteria might be relevant depending on the application. A good introduction is given in [45]. Capacity is a general term for the number of bits that can be embedded in a certain media type that is occasionally found in the literature. The more precise term throughout this book is *payload capacity*: Payload capacity expresses that only the numbers of bits that can be embedded by a user are relevant.

The actual numbers of bits used to store this payload might be much higher, such as in the case of using error correction codes. However, payload capacity is also related to other parameters. As an example, in streaming data types like audio or video, the data rate is also important. Data rate here refers to the number of bits that can be embedded in a minimum (typically temporal) segment of the carrier signal. For audio and video, a minimum segment, typically 5 sec or 1 sec is considered.

Also, the minimum carrier signal bandwidth necessary for embedding and retrieving the embedded payload correctly must be taken into account. This parameter is called *granularity*.

Since a watermarking algorithm is not always able to recover the embedded watermark, another quality metric is the error rate. This is done by analyzing the watermarking algorithm statistically or empirically. Additionally, security issues cannot be neglected. On the one hand, this involves the key capacity (the total number of keys that can be used for embedding). On the other hand, security is also derived from the detectability of the watermark or the information that can be collected by the availability of the embedder and detector. For some applications, even the complexity of the algorithm is an important criterion.

Further parameters influence the security of a watermarking algorithm and the security of a system using watermarking technology. For example, the individual implementation determines on the security of the watermarking algorithm. Using a "bad" pseudorandom number generator (PRNG) for the derivation of keying material will influence the security of the watermarking system, as will the implementation (e.g., tamper resistance of embedding and retrieval circuits as well as tamper resistance of the overall system—a defect in this may result in the elimination of watermark processing in its entirety from a protection system; see Section 9.2).

7.4.1 Concepts

Different reasons for using a watermarking benchmark suite can be identified. Typically these are:

> ‣ All watermarking algorithms have individual strengths and weaknesses that must be taken into consideration by a potential user in evaluating a given system for an application scenario.

> ‣ Watermarking system developers have an interest in judging the relative and absolute merit of new techniques or variations on existing ones. They might also be interested in detecting weaknesses for future algorithm improvements.

> ‣ Watermarking system vendors are potentially interested in an objective and independent comparison of available commercial systems, as well as in the limitations to establish boundaries for claims that may need to be justified or maintained (e.g., false positive rates).

These scenarios represent different approaches to the use of a benchmark system. Thus, different conclusions must be drawn to enable the development of a benchmark system that is able to cover all the different aspects of possible users.

> ‣ A benchmark system must have well-defined, realistic scenarios. Otherwise, neither watermarking system users nor developers will accept such an unrealistic synthetic benchmark. These scenarios are the basis for the evaluation of watermarking algorithms. For a benchmark system, this necessitates that a variety of different scenarios are provided a priori. These scenarios must be highly correlated with real world applications in terms of attacks (and their parameters), as well as in the test data (e.g., images, video, audio) used in the benchmark. Given the rapidity with which application scenarios, particularly attack mechanisms, emerge, a benchmark suite has to be dynamic and current.

> ‣ A benchmark system must be independent of developers and vendors. Ideally, a third party with no conflicting interests should have developed the benchmark system. During the development of a benchmark suite, all ideas and aspects of watermarking developers and users should have been considered as well as possible. This is of particular relevance in the case of algorithm developers and vendors, who may—deliberately or even inadvertently—be inclined to use attacks

or parametrizations that favor their specific techniques. Also, a third party should be able to perform the benchmarks in the suite under controlled circumstances and supervise the system under test to prevent alterations and manipulation. However, such implicitly trusted third parties are clearly not mandatory; some benchmarking systems can simply rely on the credibility of the users performing the evaluation.

▸ The results, either reports or certificates, must be clear and significant. A ranking score might be helpful. Unfortunately, any such ranking score will depend on the time of execution of the benchmark because of its dynamic adaptation, unless an absolute metric can be established. Therefore, time stamps (or versioning) of the test scenarios are necessary. This is important to achieve reproduceability of the results, since the benchmark suite is likely to evolve even in case absolute metrics are used, resulting in incomparable results unless versions are taken into account.

7.4.2 Automatic evaluation of watermarking algorithms

The manual evaluation of watermarking systems is time consuming. This is caused by two factors:

▸ First, the application scenario must be investigated carefully to determine the requirements for the watermarking system. This analysis is based on parameters relevant for the scenario.

▸ Second, each watermarking system that is possibly of interest must be evaluated considering these parameters. Different watermarking methods, different watermarking parameters, and a variety of attacks and their combinations must be applied on different test material.

Of course, a benchmark system simplifies this procedure by its predefined scenarios and by its automatic evaluation. Kutter and Petitcolas [36] described the automatic evaluation of watermarking systems. We will summarize the main points below.

Quality   For automatic testing, objective tests are essential. However, objective measures, which correlate well with perceived quality, do not exist for every media type.

For media types for which a good objective quality measure is not available, automatic testing results in quality evaluations which have to be considered carefully. However, testing the quality is not only relevant for the verification of the perceived quality of the watermarked data; the perceived quality of the watermarked and attacked data is also important. As a rule of thumb, a good watermarking system maximizes the perceived quality of watermarked data while minimizing the perceived quality of successfully attacked data.

Robustness   The robustness of watermarking schemes can be evaluated by applying attacks that are relevant for the application scenario under consideration. Related to this is the choice of the test data, which must also fulfill the requirements of the scenario. Robustness can be measured by considering different error characteristics like detection probability or bit error rate. However, the perceived quality has to be evaluated after performing an attack.

Attacks have to be classified for each scenario. First, nonintentional attacks, such as the production attacks of the EBU broadcasting scenario, have to be considered. Typically, these attacks are limited in their strength and can be identified easily. Second, intentional attacks are important. These attacks are not limited in their strength. Also, because of the possibility of combining different attacks, their number is almost infinite.

Not all combinations are plausible, since attackers also want to retain a certain lower bound for the perceived quality. Therefore, robustness is related to quality and vice versa. This can be expressed in a benchmark report by using graphs that represent the robustness in relation to the quality or the strength of the attack.

Capacity   As already mentioned, payload capacity or the message length is a parameter of watermarking systems. This is independent of the number of bits used for embedding the payload in the media type. Error-correcting codes will add redundancy to the information embedded, and therefore more bits are needed which may conflict with the effective channel capacity. Unfortunately, it is almost impossible to predict the capacity of an algorithm in black box tests. Other measures can partially substitute for this limitation:

  ‣ *Ratio of message length to bit error rate.* This can be tested by fixing one parameter and determining the other parameter empirically. However, determining these characteristics empirically might require extensive testing, which is equal to a very high number of embedding and detection processes.

  ‣ *Theoretical channel capacity.* The results of the previous evaluation can be used to determine how close an algorithm is to the theoretical limit.

It is, however, likely that the number of embedding and detection processes would exceed the practical limits of a benchmark suite for approximations with high precision.

Error probabilities   A watermarking system can be characterized by its detection characteristics. The following hypothesis test can be applied; possible detector outcomes are shown in Table 7.5.

H_0: The signal contains a watermark generated with the key K.
H_1: The signal does not contain a watermark generated with the key K.

▸ Detection probability $P_d = $ probability$\{D = H_0|H_0\}$: The probability of detecting an embedded watermark.

▸ False alarm probability $P_f = $ probability$\{D = H_0|H_1\}$: The probability of detecting a watermark when no watermark is embedded.

▸ False rejection (=miss) probability $P_m = $ probability$\{D = H_1|H_0\}$: The probability of not detecting a watermark, although a watermark has been embedded.

▸ Error rate $P_e = $ probability$\{$extracted message $\neq$ embedded message$\}$: The probability of detecting a different watermark than the embedded watermark.

▸ Bit error rate (BER) $P_b = $ probability$\{$ extracted bit $\neq$ embedded bit$\}$: The probability that bits flip.

However, maximizing P_d while minimizing P_f is not possible. Therefore, plotting P_d versus P_f for each watermarking system allows the comparison of different watermarking systems. These plots are well known in pattern classification as *receiver operating characteristics* (ROC) [46] and other scientific areas [47]. Two possible ROC plots are shown in Figures 7.10 and 7.11. Each point on the curve describes the characteristic relationship of the parameters shown of the watermarking system.

Table 7.5   Possible Detector Results

	H_0 *is true*	H_1 *is true*
Detector: H_0	Hit	False alarm
Detector: H_1	False rejection	Correct rejection

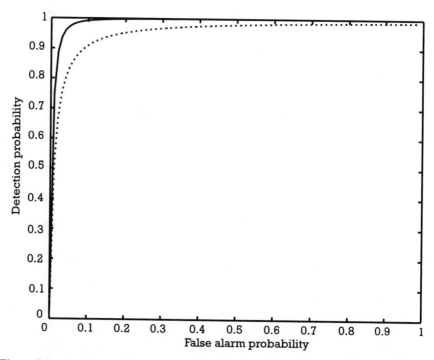

Figure 7.10 Possible ROC plot for two different watermarking systems. The detection probability versus the false alarm probability is shown. The performance of the system represented by the dotted line is worse.

Security   Kerckhoffs' desiderata [5] are essential for the development of watermarking systems. Secret keys are therefore commonly used by watermarking algorithms instead of relying on obscuring the algorithm itself. If the key payload is small, an exhaustive search might allow unauthorized detection of an embedded watermark. Also, similarities in the watermarking patterns used might result in high levels of correlation and similar problems.

Complexity   The complexity of a watermarking system might be fixed or variable for input data with the same dimension. Also, the complexity of the embedding process might be different from the complexity of the retrieval. For analyzing complexity, detailed knowledge of the system under test is necessary. For black box testing, the only way to estimate the complexity is to measure the execution time of a system. Certain scenarios require real-time processing of the watermarking system. For some scenarios, not only the real-time processing but also the latency of the system is important.

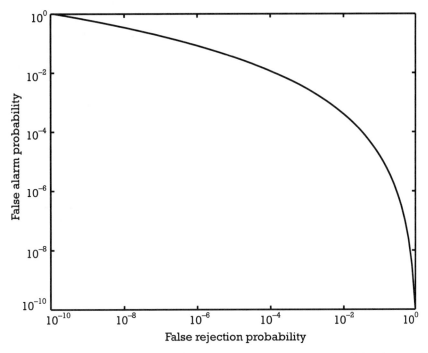

Figure 7.11   Possible ROC plot of a watermarking system considering the false alarm probability versus the false rejection probability.

Scenarios   A wide variety of scenarios is necessary for benchmarking watermarking systems with different target scenarios. These scenarios must be well defined and highly correlated with the real-world requirements. This affects the choice of attacks, their parameters, the test material, and the payload length. Additional details on the evaluation of different benchmark metrics and parameters can be found in [48].

The previous criteria are largely regarded as uncontroversial in current benchmark systems and have been applied in four benchmark systems thus far. Each benchmark stresses individual points that are important for the evaluation of watermarking algorithms. Up to version 3.1, the StirMark benchmark was the only benchmark available for the evaluation of watermarking algorithms for still images. Further developments tried to improve some weaknesses of the StirMark benchmark.

StirMark benchmark   The Stirmark benchmark as developed by Peticolas was the first benchmark that tried to analyze different still-image watermarking systems. The central concept was to test against small geometrical distortions.

A desynchronization attack in the form of a random bilinear distortion was implemented. The StirMark benchmark and its background is described in [36] and [20].

In the StirMark benchmark, a number of different attacks are applied as described in [36]. These attacks are typical image processing operations which can be applied by users of a computer system who have image processing software installed on their computers. PSNR is used as a distortion metric. The results (detected or extracted watermark) are averaged. Therefore, no explicit scenario is considered.

Currently, the StirMark benchmark aims to be a benchmark evaluation service, the "StirMark Benchmark Evaluation Service." It addresses the previously discussed issues which are important for a watermarking benchmark system. Its architecture, as described in [49], is shown in Figure 7.12.

Checkmark benchmark   The Checkmark benchmark [50] was developed by Pereira at the University of Geneva. During its design, StirMark benchmark 3.1 was state of the art and had, according to the author of Checkmark, some deficiencies. Some of the criticism was levied against the fact that only a simple quality metric (PSNR) was used. Also, prior information on the image and the watermark was not considered in the attacks. Although the StirMark benchmark contained a large number of geometrical attacks, more

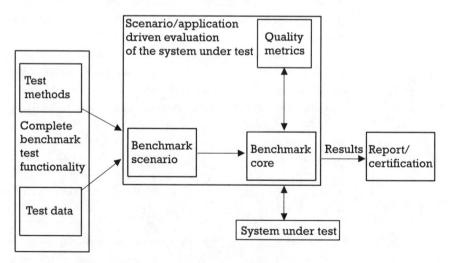

Figure 7.12   The architecture of the StirMark benchmark evaluation service [49].

sophisticated attacks were missing. Additionally, the fact that the benchmark suite was not application driven was seen as a deficiency.

Additionally, in Checkmark the Watson model of the HVS was implemented as the metric of choice. Also, projective transforms were integrated as attack scenarios. A new class of attacks was also integrated, particularly estimation-based attacks. Each implemented attack is considered as a function of the application, which allows scenario-dependent attacks.

Optimark benchmark   Similarly, the developers of the Optimark benchmark [51] recognized some deficiencies in StirMark benchmark 3.1. The empirical analysis of the results in particular was extended in the Optimark benchmark. The Optimark benchmark also takes the false alarm probability into consideration, a highly significant criterion for application scenarios that perform a large number of detection steps (especially on customer premises equipment). The nonbinary outputs are used to calculate the ROC empirically. For binary detectors, a single P_f and P_m is evaluated. Scenarios are addressed by weights used for combining the individual results.

Certimark benchmark   In contrast to the previous benchmarks, which are mainly developed by a single institution, Certimark is the result of a research project involving a broad-based consortium funded by the European Union. There are 15 partners in this consortium which range from universities, research centers, and SMEs to large corporations.

Certimark stands for Certification for Watermarking Techniques. One objective is to create a benchmark that evaluates watermarking algorithms and results in certificates. These certificates should address principal questions which initiated the usage of a benchmark.

Thus, Certimark and the StirMark Benchmark Evaluation Service are aiming in the same direction, and Figure 7.12 can also be considered as a good visualization of the underlying design principle.

References

[1] Cox, I. J., M. L. Miller, and J. A. Bloom, *Digital Watermarking*, "The Morgan Kaufmann Series in Multimedia Information and Systems," San Francisco: Morgan Kaufmann Publishers, 2002.

[2] Craver, S., B.-L. Yeo, and M. Yeung, "Technical Trials and Legal Tribulations," *Communications of the Association for Computing Machinery*, Vol. 41, No. 7, July 1998, pp. 45–54.

[3]   Hartung, F., J. K. Su, and B. Girod, "Spread Spectrum Watermarking: Malicious Attacks and Counterattacks," in P. W. Wong and E. J. Delp, (eds.), *International Conference on Security and Watermarking of Multimedia Contents*, San Jose, CA, January 1999, pp. 147–158.

[4]   Katzenbeisser, S., and F. A. P. Petitcolas, (eds.), *Information Hiding: Techniques for Steganography and Digital Watermarking*, Norwood, MA: Artech House, 2000.

[5]   Kerkhoffs, A., "La Cryptographie Militaire," *Journal des Sciences Militaires*, 9th series, January/February 1883, pp. 5–38,161–191.

[6]   Perrig, A., "A Copyright Protection Environment for Digital Images," Ph.D. thesis, École Polytechnique Fédérale de Lausanne, 1997.

[7]   International Federation of the Phonographic Industry, *Request for Proposals— Embedded Signalling Systems*, technical report, International Federation of the Phonographic Industry, London, 1997.

[8]   Su, J. K., and B. Girod, "Power-Spectrum Condition for Energy-Efficient Watermarking," *International Conference on Image Processing (ICIP-99)*, Kobe, Japan, October 1999, pp. 301–305.

[9]   Hartung, F., and B. Girod, "Digital Watermarking of Uncompressed and Compressed Video," *Signal Processing*, Vol. 66, No. 3, May 1998, pp. 283–301.

[10] Craver, S. A., et al., "Reading Between the Lines: Lessons from the SDMI Challenge," *Proceedings of the 10th USENIX Security Symposium*, Washington, D.C., August 2001.

[11] Secure Digital Music Initiative, *SDMI Portable Device Specification*, Technical Report PDWG99070802, Part 1, Version 1.0, July 1999.

[12] Langelaar, G. C., R. L. Lagendijk, and J. Biemond, "Removing Spatial Spread Spectrum Watermarks by Non-Linear Filtering," *Ninth European Signal Processing Conference*, Island of Rhodos, Greece, September 1998, pp. 2281–2284.

[13] Voloshynovskiy, S., et al., "Generalized Watermarking Attack Based on Watermark Estimation and Perceptual Remodulation," in P. W. Wong and E. J. Delp, (eds.), *Proceedings of Electronic Imaging 2000, Security and Watermarking of Multimedia Contents II*, San Jose, CA, January 2000.

[14] Cox, I. J., and J.-P. M. G. Linnartz, "Some General Methods for Tampering with Watermarks," *IEEE Journal on Selected Areas in Communications*, Vol. 16, No. 4, May 1998, pp. 587–593.

[15] Boneh, D., and J. Shaw, "Collusion-Secure Fingerprinting for Digital Data," in D. Coppersmith, (ed.), *Proceedings of Advances in Cryptology, CRYPTO '95*, Vol. 963 of *Lecture Notes in Computer Science*, Santa Barbara, CA: Springer-Verlag, August 1995, pp. 452–465.

[16] Boeuf, J., and J. P. Stern, "An Analysis of One of the SDMI Candidates," in I. S. Moskowitz, (ed.), *Information Hiding: 4th International Workshop*,

Vol. 2137 of *Lecture Notes in Computer Science*, 2001, Springer-Verlag, Pittsburgh, pp. 395–409.

[17] Linnartz, J.-P. M. G., and M. van Dijk, "Analysis of the Sensitivity Attack Against Electronic Watermarks in Images," in D. Aucsmith, (ed.), *Information Hiding: Second International Workshop*, Vol. 1525 of *Lecture Notes in Computer Science*, Portland, OR: Springer-Verlag, 1998, pp. 258–272.

[18] Kalker, T., J.-P. M. G. Linnartz, and M. van Dijk, "Watermark Estimation Through Detector Analysis," *Proceedings of the International Conference on Image Processing*, October 1998, pp. 425–429.

[19] Tewfik, A. H., and M. F. Mansour, "Secure Watermark Detection with Non-Parametric Decision Boundaries," *Proceedings of the 2002 IEEE International Conference on Acoustics, Speech, and Signal Processing (ICASSP)*, Orlando, FL, May 2002, pp. 2089–2092.

[20] Petitcolas, F. A. P., R. J. Anderson, and M. G. Kuhn, "Attacks on Copyright Marking Systems," in D. Aucsmith, (ed.), *Information Hiding: Second International Workshop*, Vol. 1525 of *Lecture Notes in Computer Science*, Portland, OR: Springer-Verlag, April 1998, pp. 218–238.

[21] Petitcolas, F. A. P., and D. Kirovski, "The Blind Pattern Matching Attack on Watermark Systems," *Proceedings 2002 IEEE International Conference on Acoustics, Speech, and Signal Processing (ICASSP)*, Orlando, FL, May 2002, pp. 3740–3743.

[22] Kutter, M., S. Voloshynovskiy, and A. Herrigel, "The Watermark Copy Attack," in P. W. Wong and E. J. Delp, (eds.), *Proceedings of Electronic Imaging 2000, Security and Watermarking of Multimedia Contents II*, January 2000, pp. 371–381.

[23] Craver, S., et al., *Can Invisible Watermarks Resolve Rightful Ownerships?* technical report 20509, IBM Research Divison, Yorktown Heights, NJ, July 1996.

[24] Cox, I. J., et al., *Secure Spread Spectrum Watermarking for Multimedia*, technical report 95-10, NEC Research Institute, 1995.

[25] International Telecommunication Union, *Methods for Subjective Assessement of Small Impairments in Audio Systems Including Multichannel Sound Systems*, 1997.

[26] International Telecommunication Union, *Subjective Assessment of Sound Quality*, 1990.

[27] Shlien, S., and G. Soulodre, "Measuring the Characteristics of 'Expert' Listeners," *Proceedings 101st Convention Audio Engineering Society*, November 1996.

[28] Precoda, K., and T. Meng, "Listener Differences in Audio Compression Evaluations," *Audio Engineering Society*, Vol. 45, No. 9, September 1997, pp. 708–715.

[29] Sporer, T., "Evaluating Small Impairments with the Mean Opinion Scale—Reliable or Just a Guess?" *Proceedings 101st Convention Audio Engineering Society*, November 1996.

[30] Zwicker, E., and H. Fastl, *Psychoacoustics: Facts and Models*, 2nd ed., Heidelberg: Springer-Verlag, 1999.

[31] International Telecommunication Union, *Method for Objective Measurements of Perceived Audio Quality (PEAQ)*, 1998.

[32] EBU Project Group B/AIM, *EBU Report on the Subjective Listening Tests of Some Commercial Internet Audio Codecs*, technical report, European Broadcasting Union (EBU), June 2000.

[33] International Telecommunication Union, *A Method for Subjective Listening Tests for Intermediate Audio Quality—Contribution from the EBU to ITU Working Party 10-11Q*, 1998.

[34] Keelan, B. W., *Handbook of Image Quality: Characterization and Prediction*, Monticello, NY: Marcel Dekker, 2002.

[35] Girod, B., "What's Wrong with Mean-Squared Error," in A. B. Watson, (ed.), *Digital Images and Human Vision*, Cambridge, MA: MIT Press, 1993, pp. 207–220.

[36] Kutter, M., and F. A. P. Petitcolas, "A Fair Benchmark for Image Watermarking Systems," in P. W. Wong and E. J. Delp, (eds.), *Proceedings of Electronic Imaging 1999, Security and Watermarking of Multimedia Contents*, San Jose, CA, January 1999, pp. 226–239.

[37] Watson, A. B., "DCT Quantization Matrices Visually Optimized for Individual Images," in J. P. Allebach and B. E. Rogowitz, (eds.), *Proceedings of Human Vision, Visual Processing, and Digital Display IV*, San Jose, CA, February 1993, pp. 202–216.

[38] Gescheider, G. A., *Psychophysics: Method, Theory and Application*, 2nd ed., Hillsdale, NJ: Lawrence Erlbaum Associates, 1985.

[39] Nadenau, M. J., "Integration of Human Color Vision Models into High Quality Image Compression," Ph.D. thesis, Signal Processing Laboratory, Swiss Federal Institute of Technology, Lausanne, November 2000.

[40] Lubin, J., "A Mathematical Theory of Communication," in E. Peli, (ed.), *Vision Models for Target Detection and Recognition: In Memory of Arthur Menendez*, Vol. 2 of *Series on Information Display*, Singapore: World Scientific, 1995, pp. 245–283.

[41] Daly, S., "The Visible Differences Predictor: An Algorithm for the Assessment of Image Fidelity," in A. B. Watson, (ed.), *Digital Images and Human Vision*, Cambridge, MA: MIT Press, 1993, pp. 179–206.

[42] Nadenau, M. J. "Integration of Human Color Vision Models into High Quality Image Compression," Ph.D. thesis No. 2296, Signal Processing Laboratory, Swiss Federal Institute of Technology, Lausanne, Switzerland, December 2000.

[43] Winkler, S., "Vision Models and Quality Metrics For Image Processing Applications," Ph.D. thesis No. 2313, Signal Processing Laboratory, Swiss Federal Institute of Technology, Lausanne, Switzerland, 2000.

[44] Video Quality Experts Group (VQEG), *Final Report from the Video Quality Experts Group on the Validation of Objective Models of Video Quality Assessment*, technical report, 2000. Available at ftp://ftp.its.bldrdoc.gov/dist/ituvidq/.

[45] Petitcolas, F. A. P., "Watermarking Schemes Evaluation," *IEEE Signal Processing*, Vol. 17, No. 5, September 2000, pp. 55–64.

[46] Duda, R. O., P. E. Hart, and D. G. Stork, *Pattern Classification*, 2nd ed., New York: John Wiley & Sons, 2000.

[47] Zweig, M. H., and G. Campbell, "Receiver Operating Characteristics (ROC) Plots: A Fundamental Evaluation Tool in Clinical Medicine," *Clinical Chemistry*, Vol. 39, No. 4, August 1993, pp. 561–577. Erratum in *Clinical Chemistry*, Vol. 39, No. 8, August 1993, p. 1589.

[48] Certimark IST 1999-10987, *D22: Benchmark Metrics and Parameters*, technical report, CERTIMARK Consortium, 2001.

[49] Petitcolas, F. A. P., et al., "A Public Automated Web-Based Evaluation Service for Watermarking Schemes: StirMark Benchmark," in P. W. Wong and E. J. Delp, (eds.), *Proceedings of Electronic Imaging 2001, Security and Watermarking of Multimedia Contents*, San Jose, CA, January 2001, pp. 22–26.

[50] Pereira, S., et al., "Second Generation Benchmarking and Application Oriented Evaluation," in I. S. Moskowitz, (ed.), *Information Hiding: 4th International Workshop*, Vol. 2137 of *Lecture Notes in Computer Science*, Pittsburgh: Springer-Verlag, October 2001, pp. 340–353.

[51] Solachidis, V., et al., "A Benchmarking Protocol for Watermarking Methods," *Proceedings of the IEEE International Conference on Image Processing (ICIP'01)*, Thessaloniki, Greece, October 2001, pp. 1023–1026.

Contents

Other content protection mechanisms

Digital watermarking by itself can offer only ex post facto evidence of violated copyright, licensing, or similar arrangements. For rights owners, a preventive mechanism is at least equally desirable; schemes for the preventive protection of intellectual property through technical means by far predate the use of digital watermarks.

8.1 Copy and usage protection schemes

The advent of general-purpose computers significantly eased the creation of copies of intellectual property; while much of the early development of software was either customized for a specific installation or so severely restricted in its possible use that unauthorized duplication and usage would have been detected easily [1], this changed as computers became mass-produced units.

The universal Turing machine's primary function of tape duplication was particularly evident once microcomputers reached a wider audience in the early 1970s—with punch tape as one of the most economical storage media available (e.g., on the MITS Altair). By then the sale of application and operating system software had become an accepted business practice; however, even with the limitations inherent in tape duplication (i.e., duplication could occur only physically; this resulted in large meetings where each individual got to duplicate the copies of others) a significant portion of the software in circulation consisted of unauthorized copies [2].

193

Software companies during the late 1970s through 1980s were concerned about losing revenue due to such piracy and implemented copy protection schemes for application programs and particularly for computer games that were at least to a significant part responsible for the success enjoyed by early mass-market microcomputers such as the Apple II and Commodore 64.

Initially, such schemes were rather simple (e.g., consisting of checking for files not immediately visible or for write protection on floppy disks) and were easily thwarted by groups and individuals that not only wanted to create pirated copies but also considered such protective measures a challenge.

One escalation strategy used by software vendors was to use unspecified or illegal data formats, such as deliberately creating files whose characteristics did not match those recorded by the directory structure of the file system and checking whether the same structure was present on a copy (commonly known as "nibble counting"), since a regular file-by-file copy would not duplicate the excess data. Similarly, the data formats for various on-disk structures were frequently altered and in some cases media were deliberately manipulated so that physical characteristics (e.g., synchronized track starts, unreadable sectors on disk, writing tracks that were inaccessible to normal disk head movements, the so-called *half-tracks*) could be verified on reading.

Within very short intervals after each such innovation in copy protection, the pirates (more commonly known as *crackers*) created countermeasures. It is a defining characteristic that the effort required for breaking a copy protection scheme had to be expended only once, after which instructions on how to circumvent a protection scheme or even ready-made tools for this purpose were made available—although in practice this was reiterated multiple times, partly due to geographically isolated groups but also due to the fact that individuals were indeed more interested in the process of cracking copy protection than in the protected material itself.

As a result, even more ambitious protection schemes such as custom recording formats fell rapidly, and countermeasures were disseminated quickly (even prior to the general use of the Internet, bulletin board systems were frequented by groups of crackers and ultimately the more general population of users), resulting not only in lost revenue due to unsold copies but also in the direct cost of creating or licensing the copy protection schemes.

In addition, many copy protection schemes relied on unspecified properties of media and devices for reading the media such as the ability to read misaligned tracks on disks or even nonexistent tracks that many devices were able to address mechanically, but were not within the performance envelope dictated for either media or devices; similarly, performance characteristics of computers and devices were used for timing-based checks, as were undocumented and hence unspecified opcodes.

This not only resulted in defects and failures as media wore out, but also precluded the use of devices that were nominally compatible with previous devices, but did not have the extraneous characteristics that copy protection scheme designers had come to rely on. This occurred regularly as, for example, manufacturers switched to improved disk drives or enhanced the performance of computers. As a result, a significant number of software products were rendered unusable for dissatisfied customers who could legitimately claim that the product as sold was defective.

Other schemes relied on the presence of additional hardware that could not be copied easily, such as the use of ROM cartridges for storing entire programs or merely small devices that could be queried for authentication and authorization (dongles). This had the obvious disadvantage of significantly increasing the marginal cost of distribution for vendors and was circumvented in a fashion similar to other protection techniques.

Somewhat similar to the use of hardware for authentication and authorization were schemes in which activation codes had to be entered on start-up or for installation; some variants included the use of *cipher wheels*, querying phrases from a book delivered with the software (the rationale being that copying a book required significantly more effort than the duplication of a digital storage medium), and code sheets printed on copy-resistant paper with color schemes that standard copiers could not reproduce and, incidentally, were extremely difficult to read.

Both hardware-based and code-based schemes (as well as the methods described earlier) were typically defeated by circumventing the protection mechanism itself. By using slightly modified computing environments up to specialized analysis tools such as in-circuit emulators (ICE), the behavior of the protection mechanism could be observed even if obfuscating techniques such as self-modifying code were used (which would defeat regular debugging tools or at least render their use more tedious) and disabled by overwriting the verification step with one that always reported a successful verification. Even sophisticated timing-dependent and self-modifying code is vulnerable to this type of attack, since ICEs operate without measurable impact on the processor under analysis and can operate at the exact speed of an unmodified environment.

Even simpler, once the program code was loaded into memory and all copy protection verification steps were executed, the in-memory image of the program could be written out with a loading mechanism that did not include the checks. While counter-counterprotection steps were attempted, run-time analysis typically defeated this type of mechanism as easily as the original copy protection mechanism.

The race between new and improved software copy protection schemes and crackers ultimately ended with the software industry all but abandoning copy protection for most software except for a small number of specialized application programs. Besides the dubious return on investment in the protection schemes, customers were frequently inconvenienced significantly by the protection mechanisms and therefore used the absence of copy protection on software as one of the purchasing criteria.

Examples for such inconveniences experienced included the inability to print due to interactions between hardware dongles attached to parallel (IEEE 1284) ports and other devices where data flowing on the IEEE 1284 interface were interpreted as command sequences for the hardware dongle, loss of copy-protected license-carrying media such as floppy disks due to wear and the subsequent unavailability of a purchased (licensed) good, and the failure of license transfer schemes (i.e., license keys had to be transferred to a legitimate installation from a license medium and returned to the license medium once the product was to be removed or transferred) due to hardware or software failures.

Even though the difficulty of surmounting ingenious copy protection systems was not inconsiderable and the number of individuals capable and interested in breaking such schemes was never large, the overall availability of cracked software was nearly ubiquitous. This was due mainly to the fact that techniques once discovered were typically encoded in automated or semiautomated programs that were then made available by direct exchange, bulletin boards, and later various Internet-based exchange mechanisms. Using such tools, even individuals with moderate to nonexistent skills were able to duplicate media or crack programs entirely.

In addition, cracked versions of programs were also made readily available, since these could be duplicated without additional skills or efforts. It is, however, interesting to note that programs for circumventing copy protection (which had legitimate applications in providing a means for backing up protected or licensed media subject to mechanical wear) were in some cases themselves copy-protected.

One of the primary underlying fallacies of the software copy protection schemes was the assumption that they had to be effective against individuals with moderate to average skills in reverse engineering and cracking (i.e., they were intended for "keeping honest people honest"). However, given that the marginal cost of distributing cracked copies or the tools for cracking was zero and that the individuals actually conducting the cracking were highly skilled and motivated not only by the challenge of breaking copy protection schemes but also by the fame and publicity in certain circles from having cracked the software and distributed it widely, copy protection schemes had to be secure not against attacks by average but by the most highly skilled individuals, even though the actual damage in terms of lost revenue was subsequently caused by the widespread use and distribution of such cracked software packages.

Moreover, the content to be protected by definition had to be available in its entirety and fully functional on a system controlled by the customer. This implied that any reverse engineering technique, up to and including mechanisms for simulating the entire run-time environment of the copy protection scheme used for verifying the integrity and validity of both itself and the content to be protected, could be used on the content. Once the protected content was decoded or otherwise activated under such observation, it could be extracted and transferred into an unprotected form.

8.2 Analog media protection

While some of the protection mechanisms discussed in the previous section relied on analog properties of the storage, in this context *analog media* refers to the recording format of the content to be protected (i.e., audio or video recordings), although some schemes also exist for print media (see above).

While there exist no protection mechanisms for analog audio media (i.e., typically reel-to-reel or compact cassette), the introduction of video recording devices in the consumer area between 1975 and 1976 and the subsequent loss of *Sony Corp. of Am. v. Universal City Studios, Inc.* (see Section 1.1) led to the prospect that devices capable of duplicating rental and home video movies would be readily available. As a result, copy protection mechanisms were introduced, the most common one from Macrovision (type I).

Analog television signals consist of 25 [in the case of phase alternation line (PAL) or *systéme electronique couleur avec memoire* (SECAM)] or 30 [National TV Standards Committee (NTSC)] frames (images) per second and 625 (PAL, SECAM) or 525 (NTSC) payload lines per frame. Frames are displayed in an interlaced format, that is, they are subdivided into two fields that are drawn on top of one another in such a way that the first field paints every other line of the frame and the second field the remaining lines required to complete the frame.

An NTSC signal consists of the following components [3, 4]:

> *Horizontal line sync pulse*   Before each line is scanned, horizontal sync pulses permit the repositioning of the electron beam to a fixed position.

> *Color reference burst*   To reach standard hue and color saturation, a 3.58-MHz color reference burst consists of a sine wave with a phase of $0°$.

> *Reference black level*   This signal provides the level corresponding to the specified maximum excursion of the luminance signal in the black direction.

> *Picture luminance information*   This signal provides the luminance, from black to peak white.

> *Color saturation*   This is interleaved with the picture luminance information on a subcarrier for backward compatibility with black and white television. The saturation of the colors is determined by the amplitude of the subcarrier. The hue of the color is determined by comparing the phase of the subcarrier with the phase of the color reference burst.

> *Color hue*   Also in the subcarrier is the color hue. The precision of color reproduction is determined by the phase of the color hue information.

> *Vertical sync pulse*   The vertical sync pulse controls the length of time of the vertical blanking interval (VBI), permitting the repositioning of the electron beam to a fixed position. This interval is also used for inserting time code, automatic color tuning, and captioning information.

Unlike most television sets, VHS video recorders compliant with the VHS patent requirements are equipped with an automatic gain control (AGC) circuit which adjusts the luminance signal by measuring the voltage difference between the bottom of the horizontal sync and the horizontal back porch (i.e., the period of time between the end of the horizontal sync pulse and the start of the next horizontal active time).

Macrovision type I copy protection for NTSC introduces false synchronization pulses within the first 40 μs in the VBI followed by false back porches at a very high voltage level (i.e., in excess of the peak white value). The AGC of a video recorder will react to this by reducing the gain to a minimal value in response to what appears to be a high amplitude signal, resulting in a low-amplitude (dark) actual signal from the video recorder or even a signal where the actual synchronization pulses can no longer be detected, resulting in "rolling" pictures. Since it would be possible to simply increase the overall gain of the resulting signal, the pulses are varied in amplitude over time.

A PAL version operates similarly, although some modifications are required due to different signal encoding [4]. This results in the image being viewable on television sets, but not on tape copies generated from such manipulated signals. Later analog protection schemes from Macrovision introduced for analog DVD reproduction also add a *colorstripe* scheme that is effective only on NTSC television sets, since it changes the color reference burst.

However, a signal distorted in this way can also lead to unstable, desynchronized images and color distortions for original video tapes or DVDs since television sets may also detect and act upon the false synchronization pulses.

The AGC protection mechanism was easily reverse-engineered by observation. Since no actual information is encoded in the VBI except for the reference black level, a trivial approach is to replace the signal with a fixed signal during the VBI; as this does not account for varying signal amplitudes between individual recordings or even within a single recording, quality typically suffers. However, by using a simple sample-and-hold circuit, the proper reference black level can be established dynamically and the VBI reconstructed from the signal thus synchronized. Such devices are sold commercially and can be built at minimal expense from readily available components by average electronics hobbyists.

Analog pay TV (broadcast via cable or satellite) was faced with similar design constraints, as the signal to be transmitted over broadcast media had to be in a valid format, but had to be accessible only to legitimate customers by means of a conditional access mechanism. The most common mechanisms for this purpose are hybrid systems in which a digital system is employed for the controlled access mechanism using cryptographic mechanisms which in turn control the scrambling of the analog signal. As with the protection mechanisms discussed in Section 8.3, all communication to conditional access enforcement devices had to be one way, since the cost of establishing communication channels for bidirectional channels would have been prohibitive. One of the more popular schemes was the Nagravision mechanism [5] developed by Kudelski, Laffely, and Sasselli; the scheme is used by a number of European pay TV broadcasters. The Nagravision mechanism performs permutations and transpositions of individual scan lines within single fields; this requires the caching of intermediate scan lines before forwarding.

The permutation function is controlled by parameters that in part reside in the conditional access mechanism at the customer's premises which can also be updated via broadcast transmission but is retained over longer periods.

The remaining part of the permutation function key is transmitted in 2-sec intervals as part of the encrypted control words transmitted over the broadcast interface.

This scrambling mechanism suffers from two main deficiencies. One is the fact that, given the long-term permutation function parameter, exhaustive search attacks are feasible with very limited computational resources. The security in this case rests on the integrity of the nonvolatile storage of the conditional access mechanism (see Section 9.2).

Beyond reverse engineering of or tampering with the conditional access mechanism itself, a more fundamental type of attack against scrambled signals (which was also employed against early voice scrambling systems) leverages properties of the video signal itself, such as the correlation between the luminosity of areas within an unscrambled field, which can occur in real time. While such correlation is not always possible immediately due to undifferentiated areas in the original video signal and can hence be disrupted if such undifferentiated areas persist over a period of time longer than a control word update, this type of attack can hardly be classified as the circumvention of an access control mechanism [6–8].

Similar video scrambling schemes, also generally vulnerable to signal processing attacks, include VideoCrypt designed by Cohen and Hashkes [9] and EuroCrypt [10].

8.2.1 Fingerprinting

As noted before, the monitoring of performances and broadcasting constitutes an application scenario of significant interest to the rights owners and becomes increasingly relevant for new types of distribution such as peer-to-peer systems. Digital watermarking permits the embedding of metainformation into multimedia content as well as the insertion of personalization information for a specific customer.

The former application requirement can, however, also be met not by embedding the metainformation within the content but rather by providing a mapping from the audio data to the actual metainformation.

The terminology for this type of approach is somewhat unsettled and confused. Besides the term *fingerprint*, occasionally qualified as *audio fingerprint* or *video fingerprint*, other descriptions refer to the mapping process as *perceptual* or *robust hashing*. The characterization of video and particularly of audio has been standardized in the form of the MPEG-7 ISO standard [11, 12]. In the context of this standard, the term *robust matching* is used. Perhaps the most confusing term is *passive* or *noninvasive watermarking*, as the marked feature here is the absence of any modification.

As with digital watermarks, the core requirements are independence of specific formats and representation, whether digital or analog, and robustness against manipulations that do not render the original signal unusable for a given application scenario. The techniques used in this case are fundamentally similar to those used in content-based identification and retrieval.

The process generally entails the extraction of salient features of the signal which can range from simple bandpass signal energy considerations [13, 14] to attempts at extracting features related to individual instruments [15] or similar feature extraction at the image level [16]. The resulting feature vector can be used by itself for identification or verification of the identity of a given signal [13, 17], or it can be used for further processing, such as the classification of signals along the lines of the MPEG-7 standard [18–20].

The main difficulty for the purposes of content protection lies in the proper balancing between robustness and the probability of false positive detection due to misclassification; for multimedia content, it is plausible that multiple distinct creations may show local or temporal similarities that can lead to a successful matching for a given creation, even though the other creations do not constitute a violation of the first creation's rights. This strongly suggests that an additional or alternate verification step is required in scenarios where the objective is the identification of copyright violations.

8.3 Off-line digital content protection

Digital content protection in off-line scenarios presents much of the same difficulties encountered with the analog content discussed in Section 8.2 while presenting a significantly greater danger, since a single instance of the protection mechanism failing alone permits, in principle, an indefinite number of identical copies.

While a number of schemes have been developed for the protection of multimedia data, the security of the schemes and veracity of vendors' claims are typically not evaluated through scientific processes. Instead, vendors frequently rely on keeping the schemes secret. This is at most relevant in the short term, since mechanisms for implementing the protection are by definition available to end users and under the complete physical control of the end user for arbitrary periods of time. The discovery of procedures and mechanisms that are deemed secret is therefore very likely to eventually fall to reverse-engineering practices.

Much as in the case of software, copy protection schemes for digital audio on CD media are retrofitted onto a system that does not support such mechanisms. While the original specification for CD players and recorders (as well as DAT, digital compact cassette, and MiniDisc devices) requires that the Serial Copy Management System (SCMS) is honored [i.e., a set of flags in the CD data indicating whether a CD may be copied at will ("copy permit"), only once ("copy once"), or not at all ("copy prohibit")], many devices simply duplicate the data unchanged including the SCMS code or are easily modified to ignore the SCMS.

One example of such a protection mechanism developed by Sony DADC is the Key2Audio family of schemes. While the original CD specification by Philips and Sony (*Red Book* [21]) specifies only audio tracks in a single session (and, correspondingly, the *Yellow Book* [22] specifies the CD-ROM representation), multiple sessions are permitted in the "Enhanced Music CD" format defined by Philips, Sony, Microsoft, and Apple (also known as *Blue Book*) which permits both audio and data sessions and specifies the directory structures for accessing the additional information. Key2Audio realizes its protection by creating a multisession CD with the audio tracks as specified in the Blue Book in the first session, and a second session containing data. However, the data track is in violation of the Blue Book standard, since the session is not finalized and is inconsistent with the directory structure.

A later variant on this, in more blatant violation of the applicable standards, was used in at least one instance of Key2Audio protection ("A New Day Has Come" by Celine Dion, Sony Music), where the table of contents claims an additional track in the audio session as the second (data) session, resulting in an illegal lead-out position. Similarly, on this particular CD the start sector for the first audio track is a negative value, which may or may not be overridden by default values in the playback device.

The desired result is that the audio session will be played back by audio-only devices, while compact disc read-only memory (CD-ROM) drives attached to computers, which can potentially be used for duplication, will not be able to recover from errors due to the inconsistent data found in the data session and will not be able to read the audio tracks.

However, a number of vendors of CD playback devices, particularly for automotive applications involving CD changer drives, use CD-ROM drives that are subject to the problems described above. This results in a legitimate claim on the part of customers that the product sold was defective, even if the CD carried the disclaimer "This CD does not play on PC/MAC," as suggested by Sony DADC. Also, older CD players that were manufactured prior to the establishment of the Blue Book do not handle multiple sessions, resulting in unspecified behavior of the device on encountering a multisession CD.

The initial Key2Audio schemes were easily defeated by obscuring the (visible) second session, such as using a felt pen. Similarly, many copying programs were able to reconstruct a Red Book–compliant audio CD copy by eliminating the extraneous data session. Even the more elaborate scheme fails if CD-ROM devices are used that provide an unprocessed data stream, which can be accomplished by modifying the field-upgradable firmware of CD-ROM devices. The resulting data stream can then be used to reconstruct a valid table of contents and audio tracks, again resulting in Red Book–compliant audio CDs.

Another example of a copy protection scheme for audio CDs, Cactus Data Shield 100, was developed by Sinquin et al. [23]. This scheme operates by inserting illegal data values instead of error-correcting codes [24] mandated by the Red Book. In addition, the values of the table of contents for the lead-out contain incorrect data.

The desired result in this case (apart from confusing CD-ROM devices honoring the content of a CD's table of contents) was to have the CD played by audio CD players, which commonly employ interpolation filters to substitute an unreadable signal (e.g., if the defect on the CD is large enough to render the error-correcting code unable to provide the additional information required for reconstructing the original signal) with an interpolation of several preceding samples until readable data can again be obtained. At the same time, CD-ROM drives, given their original design goals, typically will not emit data blocks for which error correction has failed or, in some cases, return the uncorrected data without indicating an error condition. As a result, the reading process is disrupted.

This protection scheme (and others, such as the SafeAudio scheme developed by Macrovision and TTR Technologies) suffers from the disadvantage that it eliminates a necessary element of the CD system (typically even minor disruptions such as nearby vibrations or dust particles are sufficient to cause a temporary lapse in reading the data) and deliberately causes degradation, albeit limited, even of the audio signal for which claims to the contrary are made. This is mainly due to the fact that interpolation filters can only provide adequate results if the signal remains largely static for the duration of the interpolation. This is not necessarily the case, particularly in music with significant dynamic range or where nuances introduced by individual instruments are relevant.

Moreover, error-correcting codes on Red Book–compliant CDs also provide a certain amount of protection against degrading media due to wear and tear (scratches or other blemishes such as fingerprints). This safety margin is largely eliminated, since the error-correcting and interpolation mechanisms are already employed at capacity in a new medium; any additional damage to the medium will result in a loss of data.

As in the case of Key2Audio, these techniques also result in the inability not only of using CD-ROM devices but also of other audio-specific devices employing CD-ROM drives. These problems have led to at least one class action lawsuit (*Dickey v. Universal Music Group*, filed June 11, 2002 at the Superior Court of the State of California, County of Los Angeles) against multiple vendors.

As with the software protection mechanisms discussed above, more elaborate protection schemes exploiting characteristics of "typical" devices fail on a larger proportion of devices belonging to customers, resulting not only in potentially expensive lawsuits that may well outweigh the additional revenue resulting in the reduction of the number of illegal duplications less the licensing fees required for the protection mechanism, but also in a loss of good will on the part of consumers. Moreover, as discussed above, the protection schemes can be circumvented with at most moderate effort and expenditures. While customers wishing to play back protected audio CDs on computers equipped with CD-ROM devices may employ such circumvention techniques, it appears more likely that the audio data will be distributed in the form of pirated copies once it has been extracted—even if legal protection against the circumvention of protection devices exists, this need not be the case in the jurisdiction the circumvention occurs in, and it is at least conceivable that the mere act of obtaining the audio data by circumvention can be seen as a remedy to which a consumer, having paid for the medium with an implied warranty of merchantability, is entitled to.

Therefore, vendors employing such copy protection schemes fall victim to the same fallacy discussed previously ("keeping honest people honest") for the software industry for microcomputers during the 1970s and 1980s in using techniques that did little to keep highly qualified individuals from using the content at will.

8.3.1 Cryptographic mechanisms

While the retrofit of copy protection on formats not intended for such modifications has less than satisfactory results, the design of a new format permits the introduction of more sophisticated control mechanisms. This was particularly the case with the DVD format, whose commercial success depended on the availability of major studio's movie content. The movie industry was very reluctant to release movies in high quality and particularly in digital form and hence required evidence of a significant effort to protect intellectual property in the new format.

As a result of these interactions, the DVD Video format is protected not only by technical means (discussed below) but also by a legal framework, some of which is not available publicly. Any medium on which DVD Video content is to be prerecorded must be licensed, and the same holds true for playback devices. The licensing agreements tie the ability to play back DVD Video to the inclusion of several copy protection mechanisms as well as another revenue-enhancing mechanism for the movie industry in the form of region codes.

Region codes subdivide the world into six zones plus one for special international venues (e.g., aircraft, cruise ships) and are encoded on the DVD for the target market. A compliant playback device must itself have an embedded region code and is only permitted to play back either disks matching its own region code or code-free disks (i.e., for which no region code was set). This schema was intended to permit the staggered release of new movies to different regional markets without imports from other markets (e.g., causing DVDs to appear at the same time as the movie played in cinemas).

However, since vendors had to keep the cost of adapting devices to a given market at a minimum, techniques for switching regions without hardware modifications or requiring only minimal modifications were typically employed, such as command sequences to be entered via the infrared control port or by modifying the firmware of a device.

Some vendors permitted their playback devices to be adjusted in such a way that they did not verify the region code at all, thereby accepting all regions equally. Several movie studios reacted to this violation of the licensing agreement by introducing active checks for a region code matching that on the DVD Video medium and setting all region code flags active; this technique was originally introduced by Warner Bros. in late 2000 as "Region Code Enhancement." While this thwarts code-free playback devices and autoswitching devices (i.e., devices recognizing the region code for a given disk and adjusting the player region code automatically), the latter cannot use the region code flags to determine the proper region, a problem that also affects newer DVD-ROM drives which are manufactured without a preset region code and adjust to the region code of the first media encountered. Even this enhanced scheme, however, does not provide protection against manually switched region codes.

Another protection mechanism for analog output that is part of the DVD Video specification—although its use is not mandatory due to licensing fee considerations—is the Macrovision AGC mechanism (optionally including the colorstripe mechanism where the signal norm permits), discussed in Section 8.2.

Analogous to the SCMS for audio CDs is a serial copy management system called Copy Generation Management System (CGMS) which is embedded in the analog and digital signals and may be honored by devices such as camcorders.

However, the main protection afforded to DVD Video comes in the form of an integrated encryption scheme, content scrambling system (CSS). CSS was developed primarily by Matsushita and Toshiba (licensing is administered by the DVD Copy Control Association) and consists of several components. The actual content is encrypted using a stream cipher that was kept confidential in violation of Kerckhoffs' principle, and whose key is subsequently encrypted by a number of master keys that are stored on each DVD. Each disk contains a 5-byte hash value of the decrypted disk-specific key for verification purposes and up to 409 encryptions of the disk-specific key.

A device must authenticate itself to the drive to gain access to locked sectors to obtain disk keys and title keys required for actual decryption of the video data (which are derived from the disk key). The data can also optionally be decrypted by a key derived from the title key by an exclusive-or of specified bytes from the unencrypted first 128 bytes of the (2,048-byte) sector. The existence of a master key list permits the removal of a licensee's key if the licensee has been found in violation of the license agreement on subsequent disks; however, such a key revocation affects only disks manufactured after the license violation was detected.

As with the noncryptographic schemes discussed earlier, a number of vulnerabilities were quickly found. As there is no mechanism to detect duplicates, large-scale pirates were still able to produce exact duplicates of DVDs once the necessary devices became available; a bitwise copy of an existing disk is not affected by any of the protection mechanisms discussed above.

The simplest scheme besides actual bitwise copying employed by pirates was the extraction of the digital signal after it had been decrypted by a legitimate playback device. Initially limited to modified hardware of playback devices (i.e., tapping data buses and extracting the MPEG-2 data stream) and therefore requiring significant expenditures, this was changed by the advent of software-based playback mechanisms that could be reverse-engineered efficiently to yield the decrypted or decoded media stream. As in the case of software copy protection, there is no protection against such an attack unless tampering can be ruled out, since both authentication and other protection mechanisms including encryption schemes are circumvented.

The CSS algorithm was originally intended for deployment in customized hardware, as at the time of the original development, decryption at the data rates required for the decryption of the MPEG-2 streams contained in a DVD was unsuitable for software implementations on general-purpose microprocessors. Since software-based playback had become desirable at a later stage during the design process, the cipher was weakened to lower the computational complexity to a level deemed acceptable. In addition, the key length was restricted to 40 bits so as not to endanger the exportability under the strict Japanese export regime for cryptographic mechanisms in place at the time.

The putative CSS cipher is a simple dual linear feedback shift register (LFSR) stream cipher, which is used for both authentication and the decryption of disk keys, title keys, and data blocks based on minor variations of the concatenation of the LFSRs. Although brute force attacks on a 40-bit cipher can be executed with relative ease, a number of attacks on CSS were demonstrated by Stevenson [25] which, depending on whether plain text is known, can result in key recovery with either 2^{16} or 2^{25} operations; both attacks can be performed in real time or require at most a few seconds.

However, even this computational effort was not necessary, since it was possible to reverse-engineer one of the master keys from a software-based implementation; as a result, the vendor's master key could be used to decrypt arbitrary CSS-protected content.

While the key list permitted a measure of renewability for CSS, the removal of the reverse-engineered key would have had only a limited effect, since knowledge of one key permits the recovery of other keys through various (brute force or cryptanalytical) means. As a result, once the initial key was published, the CSS system was defeated in its entirety. Attempts by the DVD Copy Control Association to bar the publication under trade secret legislation was rejected on appeal (*DVD Copy Control Association v. Andrew Bunner*, H021153; California Court of Appeal, 6th Appellate District, November 1, 2001), a decision that was based on a precedent dating back to strong export controls for ciphers. (In *Junger v. Daley*, 209 F.3d 481, United States Court of Appeals for the 6th Circuit, 2000 source code was ruled to be protected as speech under the First Amendment of the U.S. Constitution.)

As a result, the effectiveness of the CSS system was reduced significantly even though licensees are still required to apply CSS protection to their content. Attempts at introducing an enhanced version of CSS (CSS2) were abandoned due to concerns regarding the technical viability of such protective measures.

However, in the form of national legislation based on the WIPO Copyright Treaty (especially Article 11) and the WIPO Performances and Phonograms Treaty (especially Article 18), such as the U.S. Digital Millennium Copyright Act (DMCA), a renewed attempt was made to augment the a priori insufficient technical means of protection by additional legal means. In an extreme interpretation, this would appear to render the technical merit of a technical protection system irrelevant, since the very fact of its existence would be sufficient to also confer legal protection [26].

References

[1] Haigh, T., "Software in the 1960s as Concept, Service, and Product," *IEEE Annals of the History of Computing*, Vol. 24, No. 1, January 2002, pp. 5–13.

[2] Gates, III, W. H., "An Open Letter to Hobbyists," *Computer Notes*, December 1976.

[3] International Telecommunication Union, *CCIR Recommendation 601-2: Encoding Parameters of Digital Television for Studios, Recommendations of the CCIR*, Vol. 10, Part 1: "Broadcasting Service (Television)," 1990.

[4] Benson, K. B., and J. C. Whitaker, (eds.), *Standard Handbook of Video and Television Engineering*, 3rd ed., New York: McGraw-Hill, 2000.

[5] Kudelski, A., L. Laffely, and M. Sasselli, "Système de télévision à péage," European Patent 0 626 793, April 1987. Granted in November 1994.

[6] Mangulis, V., "Security of a Popular Scrambling Scheme for TV Pictures," *RCA Review*, Vol. 41, No. 3, September 1980, pp. 423–432.

[7] Raychaudhuri, D., and L. Schiff, "Unauthorized Descrambling of a Random Line Inversion Scrambled TV Signal," *IEEE Transactions on Communications*, Vol. 31, No. 6, June 1983, pp. 816–821.

[8] Kuhn, M. G., *Analysis of the Nagravision Video Scrambling Method*, technical report, University of Cambridge Computer Laboratory, Cambridge, U.K., August 1998.

[9] Cohen, M., and J. Hashkes, "A System for Controlling Access to Broadcast Transmissions," European Patent 0 428 252, September 1990. Granted in November 1997.

[10] European Committee for Electrotechnical Standardization (CENELEC), *Access Control System for the MAC/Packet Family: EuroCrypt*, EN 50094, 1992.

[11] ISO/IEC Joint Technical Committee 1 Subcommittee 29 Working Group 11: Coding of Moving Pictures and Audio, *Information Technology—Multimedia Content Description Interface*, ISO/IEC 15938, 2002. Published in eight parts.

[12] ISO/IEC Joint Technical Committee 1 Subcommittee 29 Working Group 11: Coding of Moving Pictures and Audio, *N4674: MPEG-7 Overview*, 1995.

[13] Haitsma, J., T. Kalker, and J. Oostveen, "Robust Audio Hashing for Content Identification," *Proceedings of the Second International Workshop on Content Based Multimedia and Indexing*, Brescia, Italy, September 2001.

[14] Papaodysseus, C., et al., "A New Approach to the Automatic Recognition of Musical Recordings," *Journal of the Audio Engineering Society*, Vol. 49, No. 1, 2001.

[15]   Brown, J. C., "Computer Identification of Musical Instruments Using Pattern Recognition with Cepstral Coefficients as Features," *Journal of the Acoustical Society of America*, Vol. 105, No. 3, March 1999, pp. 1933–1941.

[16]   Kivanç Mihçac, M., and R. Venkatesan, "New Iterative Geometric Methods for Robust Perceptual Image Hashing," in T. Sander, (ed.), *Security and Privacy in Digital Rights Management: ACM CCS-8 Workshop DRM 2001*, Vol. 2320 of *Lecture Notes in Computer Science*, Philadelphia: Springer-Verlag, November 2002, pp. 13–21.

[17]   Kalker, T., "Applications and Challenges for Audio Fingerprinting," *Proceedings of the 111th Audio Engineering Society Convention*, New York, December 2001.

[18]   Allamanche, E., et al., "AudioID: Towards Content-Based Identification of Audio Material," *Proceedings of the 110th Audio Engineering Society Convention*, Amsterdam, May 2001.

[19]   Hellmuth, O., et al., "Advanced Audio Identification Using MPEG-7 Content Description," *Proceedings of the 111th Audio Engineering Society Convention*, New York, December 2001.

[20]   Kastner, T., et al., "MPEG-7 Scalable Robust Audio Fingerprinting," *Proceedings of the 112th Audio Engineering Society Convention*, Munich, May 2002.

[21]   IEC, "Compact Disc Digital Audio System," IEC 908, 1995.

[22]   ISO/IEC Joint Technical Committee 1 Subcommittee 23: Optical Disk Cartridges for Information Interchange, *Read-Only 120 mm Optical Data Disks (CD-ROM)*, ISO/IEC 10149, 1995.

[23]   Sinquin, P., P. Selve, and R. Alcalay, "Anti-Counterfeit Compact Disc," United States Patent 6,208,598, January 1999. Granted in March 2001.

[24]   Lin, S., and D. J. Costello, Jr., (eds.), *Error Control Coding: Fundamentals and Applications*, Prentice-Hall Series in Computer Applications in Electrical Engineering, Englewood Cliffs, NJ: Prentice Hall, 1983.

[25]   Stevenson, F. A., "Cryptanalysis of Contents Scrambling System," Usenet posting, November 1999.

[26]   Marks, D. S., and B. H. Turnbull, " Technical Protection Measures: The Intersection of Technology, Law and Commercial Licenses," Workshop on Implementation Issues of the WIPO Copyright Treaty (WCT) and the WIPO Performances and Phonograms Treaty (WPPT): World Intellectual Property Organization, December 1999.

Integrated content protection solutions

The protection mechanisms discussed in Chapter 8 were mainly targeted at copy protection and conditional access for broadcast media. Digital representations, however, provide a significant number of possible usage scenarios in which the ability to use (e.g., view or hear) a creation is concomitant with the ability to duplicate and redistribute or even modify the creation; the operations necessary for the primary purpose are, at a sufficiently abstract level, indistinguishable from those for duplication and redistribution, since digital representations need to be copied and otherwise processed for its intended purpose as well.

The obvious conclusion from the discussion so far is that it is imperative to protect creations along the entire digital but also analog processing chain and to ensure that the intellectual property rights to the creations are upheld. This necessitates the integrated and interlocking use of security and protection mechanisms for storage, transfer, and evaluation of multimedia creations, frequently summarized under the heading of *digital rights management* (DRM).

9.1 Digital rights management

The term DRM is rather ill-defined and has been used almost synonymously with content protection on one end of the spectrum and also in describing specific technical implementation mechanisms. The following provides a number of general objectives for a DRM system; these are posed by several interested parties and are to some extent mutually exclusive:

> ▶ Intellectual property rights throughout the distribution and dissemination chain must be protected.

> ▶ End-user interests must be protected (e.g., in the case of DRM technology provider failure, the rights of end users to the content they purchased need to be maintained—both in the face of technical and business failures).

> ▶ There must exist a facility for specifying the precise rules and regulations to be applied to the usage of creations. This implies a clear separation between the content and the rights, as the same content can be licensed or sold under multiple different conditions.

> ▶ There must exist a clear separation between the mechanism for identifying rights to creations and the mechanism for enforcing them. This is in recognition of the fact that both the underlying distribution and protection mechanisms layered on top of the distribution mechanisms will change over time, but contractual obligations generally do not and need to be enforced identically even if such changes occur.

> ▶ The impact in terms of inconveniences incurred on the part of end users must be negligible.

> ▶ Development, unit, and administrative costs of the DRM mechanism must be minimized and balanced against actual losses incurred by flaws or limitations within the DRM.

> ▶ The system must be renewable, that is, in the event that a device, user, or vendor is partially or completely compromised, the compromised entities must be deactivated or kept from causing further damage while retaining compatibility with existing, noncompromised entities. Furthermore, migration of existing systems to new protection mechanisms while retaining backward compatibility is necessary.

The protection requirements can also be grouped approximately into several categories:

> ▶ The basic premise of any DRM system is the provision of access control based on well-specified rights. The latter requirement, while theoretically within the realm of basic functionality of operating systems, is typically not met by these due to both a lack of expressiveness in formulating the rights and enforcement mechanisms generally limited to discretionary access controls unsuitable for denying users access to resources. Depending on the usage scenario, this is accompanied by a

need for proper identification and authentication as a prerequisite for access control decisions.

▸ Beyond access control, it may be necessary or desirable to further restrict the usage of resources (i.e., creations) or certain aspects thereof. Such usage can, for example, consist of a restriction on the absolute time frame or duration of usage, the number of times a resource is used, or certain aspects and selective components of a creation such as a foreign language audio track in a motion picture or a high-resolution three-dimensional model suitable not only for on-line viewing but also for stereolithographic reproduction.

▸ Protection of persistent information against manipulation and interception, whether access control information such as key or rule material or the creations themselves, must be protected in transit (regardless of the form of delivery, for example, physical media such as DVDs and point-to-point, point-to-multipoint, or broadcast transmissions) and storage.

Particularly for the protection of the creations, this implicitly includes protection against redistribution in a form that is usable to unauthorized recipients. In addition, the requirements for the protection against manipulation can be further elaborated to include the prevention or detection of distribution or use of unauthorized modifications, as well as the mandatory inclusion of certain inband information such as visible copyright notices and usage restrictions.

In addition, it can be desirable that the protection mechanisms extend to fragments of protected creations (e.g., a cropped image), collections of creations in which at least one protected creation is contained (e.g., scenes created from multiple three-dimensional models), or both.

▸ Identification, authentication, and authorization of rights granting and revocation are required, particularly for content stored or received by an end user or device and transient granting mechanisms. These requirements exist both in transient (interactive) and persistent forms and also extend to the requirement to identify and persistently prove the delivery of creations and the association of a licensee or end user with a specific instance. Similar identification and authentication may also apply to the creations themselves, necessitating fingerprinting techniques (see Section 8.2.1) of the creations for such applications.

▸ Traceability or trackability of creation instances or transactions leading to rights granting and transfer may also be required; whereas the tracing of transactions occurs by definition in the digital domain, transfers requiring tracing or tracking can occur both in digital and analog domains. This implies a need for techniques robust against such conversions between digital and analog and similar manipulations.

This may also apply, as above, to excerpts, collections, or both consisting of protected creations.

› Anonymity or pseudonymity of transactions may constitute a requirement; that is, the subject of transactions such as which creation (e.g., an electronic book) was transferred or to which rights are granted should not be associable with the individual (in the case of anonymity) or not immediately associable by the provider of the creation or other entities such as law enforcement authorities unless cause is demonstrated to a third party holding identities in escrow (pseudonymity).

› The rights to creations should not be tied to individual hardware or other devices, as such devices may be corrupted, destroyed, or obsolete, whereas the rights need not be subject to the same degradation.

As with the overall requirements, the protection requirements in some cases are contradictory, mainly based on the origin of the requirement (e.g., the rights owner's desire for traceability versus the requirement for anonymity or pseudonymity on the part of end users).

A simplified and idealized DRM system model is shown in Figure 9.1. Rights owners process their material into a form determined by the DRM

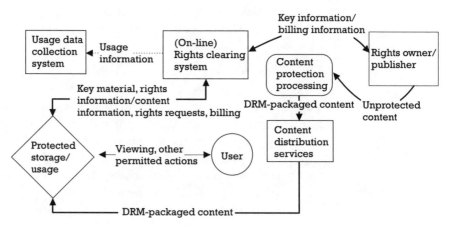

Figure 9.1 Idealized DRM system model.

system and provide rights markup to determine the conditions, pricing, and usage types associated with their creations. This typically is accomplished by encrypting the creation, associating a unique identifier [1] with it and embedding the resulting data in a well-defined format (e.g., [2, 3]).

Such data are then distributed either via explicit content distribution services (e.g., multimedia streaming, web services) or "superdistribution," that is, distribution via other users to the end users or, more specifically, to one or more protected storage/usage units which serve to play back or otherwise facilitate usage of the creations or, in the case of persistently stored creations, also serve as a storage facility enforcing the DRM-specified rights and protecting the stored data as well as the protection mechanism itself against tampering and extraction.

In the case of usage, the DRM system can—either locally within the protected storage/usage units or in collaboration with an interactive service—then evaluate the rights granted to the user against the desired operation and, if necessary, inquire for additional services, as well as initiate payment (unless such has not already been rendered in the course of distribution). Assuming rights are granted, a storage/usage unit can perform the desired operation. In the case of an interactive rights clearing system, the various information collected (which may include, but is not limited to, the user's identity, device identities, content identification, time of usage, and type of usage) can also be gathered by a separate usage data collection system for further processing.

The following sections briefly discuss technology examples that illustrate both the approaches used and the challenges faced by DRM mechanisms that attempt to provide seamless, transparent end-to-end protection given the constraints of the target application scenario. It should be noted, however, that to the best knowledge of the authors, no dissemination or full disclosure of DRM systems in the scientific literature exist (one of the present volume's author's implementation described in Section 9.1.3 notwithstanding, since that system is based on different protection and usage assumptions); existing DRM systems primarily rely on the secrecy of implementation mechanism for the protection they afford.

Section 9.1.1 discusses a specific protocol (DTCP) for point-to-point copy control and management between federated devices. The important aspect of integrity protection and assessment and its implications for the realization of digital rights management is then exemplified in an architectural specification (TCPA) in Section 9.1.2, while Section 9.1.3 discusses the prerequisites and implementaton of a sample content protection architecture intended for commercial, benign environments.

9.1.1 The Digital Transmission Content Protection Specification

While it is desirable both from users' perspective in terms of functionality, quality, and potential convenience and device manufacturers' perspective in terms of ease and cost of implementation to use digital interconnection and transmission formats for consumer electronics devices, any protection or full DRM mechanism must therefore ensure that the requisite protection is not only afforded within an individual device or computer system, but by any and all other devices receiving the digital content.

The goal of the Digital Transmission Content Protection Specification (DTCP) as laid out by the five principals, Hitachi Ltd., Intel Corporation, Matsushita Electric Industrial Co. Ltd., Sony Corporation, and Toshiba Corporation (also known as the 5C) upon founding in February 1998 was therefore to "stop unauthorized, casual copying of commercial entertainment content" [4] based on the assumption that the devices interconnected by means of DTCP enforce the DRM requirements individually and communicate only via DTCP or another protocol that enforces the relevant rights specifications in conjunction with the requisite identification, authentication, and authorization.

The DTCP system has been approved for export by Japan's Ministry of International Trade and Industry (MITI), which presumably in part explains the ciphers used. It is primarily designed for consumer electronics devices using the IEEE 1394 serial bus but can also be used on appropriately equipped PC systems and other digital interconnection systems such as PCI and Universal Serial Bus (USB) [5].

Conceptually, protection of DTCP is modeled after the SCMS found in earlier digital recording devices (see Section 8.3) but it provides several levels of allowed copying carried in the *copy control information* (CCI), namely:

Copy Free   The content can be copied arbitrarily.

Copy Once   The content can be copied exactly once.

Copy No More   The content has been copied before and is not to be copied again.

Copy Never   No copies of the content can be made.

This CCI is transmitted together with (but not embedded in) the content that is encrypted by the source device using a symmetric cipher; the 56-bit key length M6 block cipher was selected as the baseline; other ciphers may optionally be supported. Authentication is achieved either by means of public key mechanisms in the case of Full Authentication by exchanging signed device certificates, random challenges, and cipher key components (this variant supports all copy mechanisms) or by secret key mechanisms in the case of Restricted Authentication, which supports only Copy Once and Copy No More.

There are three components for copy protection in the DTCP specification in addition to the CCI:

Device authentication and key exchange   This mechanism ensures that connected devices are able to verify the authenticity of other devices. There are two modes for authentication, *Full Authentication* and *Restricted Authentication*. The former supports all policies, the latter only Copy Once and Copy No More. Each device (which may also be a general-purpose computer) can only act as either sink or source device for data transfers. The data transfers themselves will occur only after a successful authentication.

Content encryption   The standard requires devices to have a cryptographic subsystem capable of encrypting and decrypting exchanged data using at least the baseline cipher (the M6 block cipher).

System renewability   Devices that support Full Authentication must support certificate revocation lists. These are called *System Renewability Messages*. The certificates used by DTCP are completely proprietary and identify units with a 40-bit device ID. This obviously raises issues of privacy. Devices are assigned a unique device ID (X_{ID}) and a device Elliptic Curve Digital Signature Algorithm (EC-DSA) asymmetric key pair (X^1, X^{-1}) by the Digital Transmission License Administrator (DTLA). X^{-1} must be stored persistently in the device; compliant devices are also assigned a device certificate X_{Cert} by the DTLA which is also stored in the device.

The algorithms used by DTCP for authentication are elliptic curve cryptography using the parameters in [6]; the hash function is SHA-1 [7].

While the cryptographic mechanisms for authentication can be considered adequate, the choice of the proprietary M6 as bulk encryption cipher permits not only brute force attacks as were conducted against the DES cipher [8, 9], but it is also vulnerable against specific classes of attacks [10]. Other protection solutions for point-to-point connections, specifically the High-Bandwidth Digital Content Protection (HDCP) protocol developed by Intel Corporation for the cryptographic protection of video data for immediate display purposes using the digital video interface (DVI) [11], have not fared significantly better; in fact, for HDCP, multiple vulnerabilities were discovered [12].

The most severe weakness besides the risk of tampering, however, concerns the renewability mechanism. DTCP contains a certificate revocation mechanism, called *system renewability message* (SRM). There exists a special (extensible by means of a *generation* field) format for SR messages; however, the first-generation SRM definition is rather simple and consists merely of a header followed by a number of individually signed certificate revocation list (CRL) parts which are subsequently signed as a whole. All signatures are provided by the DTLA (the licensing administration body controlling the overall integrity and renewability) and are 320 bits long. The SRM contains a 16-bit version number (SRMV) which is to increase monotonically, limiting the total number of updates possible with this scheme to 65,535 instances, since it is not reset upon generation change. The total length of the CRL is also indicated by a 16-bit field, imposing a severe limit on the total number of certificates (and thereby devices) that can be blacklisted, considering that one can either blacklist individual devices or blocks of up to 65,535 devices. This severe restriction may be alleviated by introducing further generations of devices, but considering the fact that consumer devices are the target of the DTCP specification, it seems inevitable that backward compatibility exerts a strong influence toward forgoing deactivation of older equipment that cannot support the new SRM generations.

SRM exchanges can occur whenever compliant devices are connected by including SRM on prerecorded media or with digital streaming media servers. The device is supposed to verify the SRM version number and, if the version number of the received SRM is larger than the one stored locally and the digital signature on the SRM is verified, to store as much as possible locally; the device is not required to store the entire SRM. This opens the possibility of attacks in that devices compromised after a certain point in time communicating with other devices that do not have adequate storage space for SRM messages cannot be detected.

9.1.2   The Trusted Computing Platform Alliance System architecture

The Trusted Computing Platform Alliance (TCPA) was initially formed by Compaq Computer Corporation, Hewlett-Packard Company, IBM Corporation, Intel Corporation, and Microsoft Corporation.

The TCPA specification specifies a trusted subsystem or coprocessor as an integral, nonbypassable part of each platform and defines interfaces for operating systems, applications, and external parties for accessing integrity metrics.

The integrity metrics provided by the trusted platform module (TPM) ensure, for example, that the software being executed is accredited as trusted or that the overall system configuration meets requirements set forth for the execution of further trustworthy code or the release of certain data material [13].

For this purpose, a single public key infrastructure certification authority assigns each TPM (which in turn is associated with precisely one platform, i.e., a device or computer) an identity and the cryptographic means (via public key cryptography) to prove the identity to third parties also trusting the TCPA certification authority. This, in conjunction with an additional public key digital signature asserting the conformance of the given platform with a valid and current TCPA specification, permits the identification and authentication of conforming and protected platforms. Maintainability (or *renewability* in DRM parlance) is provided by the ability to update the key material and authentication information stored by a TPM.

Upon initialization, the TPM must ensure that no programs other than those intended by the entity that vouches for the root of trust for measuring integrity metrics are executed, must accurately measure at least one integrity metric that indicates the software environment of a platform, and must accurately record measured integrity metrics in a designated storage area within the TPM.

Once such integrity metrics are gathered, these can be queried by a trusted platform agent (TPA) in the form of a challenge that yields the measurement results along with validation data that provides a metric for verifying the integrity metric data.

One of the core applications of the TCPA is to permit the *sealing* or *binding* of certain software or data to a platform for which trustworthiness has been endorsed by cryptographic means. Such software may be an operating system or plain payload data, since the actual data are opaque to such operations.

It is precisely the sealing mechanism described above that has led to the assumption that the TCPA is intended for the implementation of DRM mechanisms (i.e., as storage units described in Section 9.1 and, by extension in the execution of well-defined—trusted—code, also as usage units) [14, 15], although this is explicitly denied in a document released after severe criticism was raised in both academic and general publications [16]. While the statement in [16] denies that the primary goal of the TCPA is usage as a DRM (and indeed, the TCPA provides valuable services for a number of applications, particularly in the creation of high-assurance operating systems), this statement, however, cannot be entirely discounted.

Since the TCPA specification document explicitly does not elaborate on the mechanisms for achieving the requisite tamper resistance and other requirements such as the quality of random or pseudorandom numbers, the level of assurance that can be placed in the individual implementations may vary significantly. This implies that each individual implementation must be evaluated independently (as well as against a protection profile or security target [17–19]) to ascertain the functions and assurance provided (see Section 9.2). However, integration of TPM functionality into central processing units (CPU) manufactured at state-of-the-art feature sizes and possibly utilizing the considerable complexity afforded such CPU designs as decoy and deception tools offers significant improvements in assurance over, for example, the dongle mechanisms discussed in Section 8.1. Based on the assumption that the TCPA is indeed intended for DRM purposes (i.e., to be integrated with the actual DRM and supporting mechanisms in both operating systems and application programs), the criticism in [14, 15] is aimed firstly at the inherent lack of privacy necessitated—even in the presence of a trusted third party holding pseudonyms in escrow—by the unique endorsement key of a TPM and hence platform. The sole protection in this scheme is indeed the fact that the trusted third party will not release information or collude with the certification authorities. While acceptable in commercial environments, such schemes appear problematic for end-user applications where expectations (or legal requirements) of privacy must be taken into consideration. In addition, any DRM system relying on the endorsement key as identification of the DRM must not naively create a bijection between the TPM and an individual or even usage rights of that individual. In such a naive system, not only would this potentially limit the usage rights to a specific device (or *platform* in TCPA terminology), a restriction that is already in place in several existing DRM systems and that may infringe on the rights of the customer and violate implicit warranties of merchantability, but would also potentially link multiple individuals to the same TPM (e.g., in case a device is resold by an individual), potentially exposing later owners of a given TPM to incriminations based on the DRM usage history. While the implementation of a mechanism based on public key cryptography is both straightforward and efficient, an approach more suitable for application outside of commercial or similar environments without expectations of privacy would have considered the use, for example, of zero-knowledge proof systems or similar provable anonymizing mechanisms for achieving the required level of authentication even at a considerable cost in computational or communication complexity for users requiring such services.

Another criticism leveled against the TCPA must be considered more a criticism of potential usage of the infrastructure provided by the TCPA under the assumption that the requisite hardware mechanisms achieve broad dissemination. The existence of a sole certification authority capable of determining which data (i.e., application program or operating system) may be used on a system with the TPM enabled (which assumes that the TPM must be made operational as is, for example, the case if a single program requires the services of the TCPA) implies that it is hence also able to either deny the requisite endorsement or significantly increase the effort required to obtain proper endorsement for having such code operational with TCPA active. Besides the creation of a mechanism for multiple certification authorities that can be established without the approval of a single instance, such issues cannot be dealt with entirely at the technical level. Rather, organizational safeguards are required to alleviate the concerns of individuals or other entities that consider an oligopoly or even monopoly potentially controlling significant portions of the software executed worldwide a threat.

9.1.3 The CIPRESS system

The CIPRESS[1] system developed by one of the authors on behalf of the Mitsubishi Corporation can be considered a DRM system for commercial application areas. The distinction from other DRM systems intended primarily for the protection against end users of multimedia entertainment products is relevant in that it defines both the underlying risk analysis and provides a rationale for the implementation strategy chosen.

In the application scenario, it must be ensured that intellectual property (e.g., research and development results and competitive intelligence) as well as other information (e.g., confidential material regarding commercial or other financial transactions) is retained confidentially within the confines of a well-defined secured (corporate) area or network which may extend over a large number of network elements distributed worldwide. Within this closed user group, however, discretionary access control mechanisms are considered adequate, as the overhead in terms of inefficiencies and personnel cost associated with mandatory access control mechanisms is unattractive. Instead, the application scenario not only required to restrict the egress of material from the closed user group to closely audited channels but also to be able to control and track the usage of documents or other creations and possible derived data, as well as to protect and verify the integrity of designated data.

Another important element of the application scenario is that the (heterogeneous) systems to which the protection mechanism must be applied consist of a broad range of application programs, most of which are commercial off-the-shelf (COTS) or customized applications that cannot be modified to use DRM-specific protection mechanisms such as customized application programming interfaces. Lastly, two additional fundamental assumptions of the system are that, firstly, the users of the system operate the COTS

---

1. Cryptographic Intellectual Property Rights Enforcement System.

systems without administrative privilege or the means to elevate their privileges on the system through existing applications, and secondly, that in the environment the systems operate, tampering with computer systems is prevented through organizational safeguards and sensors that deny potential attackers access to the protected systems through channels other than software interfaces.

Based on these assumptions, CIPRESS adds security mechanisms at the kernel level directly as well as indirectly through interposition of protected nonkernel code and is thus able to enforce the DRM asset management and tracking policies for all applications and users while being largely invisible to users and applications; the kernel/user mode separation is also used for protecting itself; similar mechanisms have been used earlier [20, 21] to add nonstandard functionality to existing operating systems [22].

The enforcement of the security policy for storage media and network traffic can be ensured through the use of encryption that is interposed transparently in such a way that data are mandatorily encrypted and automatically decrypted without the possibility of user intervention, while the process of decryption is tied to a user or application process both being inside a closed user group operating environment and having the authorization for accessing or otherwise using a datum. The use of encryption particularly for all storage media as an enforcement mechanism also ensures that the DRM policy is enforced even if the DRM itself is not active.

The requisite key material is not stored persistently on end (client) systems but rather forwarded from a trusted site to a trusted environment (while CIPRESS supports hardware extensions for this purpose, a software-only implementation is typically used based on the risk analysis for a given client system) on the client system wishing to perform the storage or load/usage operations. Since verifying each file system access via a centralized database would be highly impractical, distinctions are made between general data (tied to a specific client system, this is also accomplished persistently by means of encryption) and creations intended for exchanging between individual systems or users. Only for the latter so-called *registered documents* the central access and use control mechanism is used. After users have identified and authenticated themselves to the central DRM system (in the form of a single sign-on process), all operations for which the users or processes and applications acting on behalf of the users are authorized are performed transparently and without further user interaction.

Creations that are identified for transfer to other systems (i.e., which are not locked to a specific client) are identified by a cryptographic hash value that is affixed as a label together with other administrative information to the creation. This occurs invisibly to application programs or users, as the labeling, delabeling, and other processing (particularly cryptographic operations) occur transparently.

CIPRESS is not a multilevel security system in the traditional sense; it merely enforces the access and use control mechanisms on designated creations. The remaining component of the security concept that facilitates the move beyond rigid compartmented levels for the application scenario is that of tainting. Files created or merely touched by users with write access are automatically encrypted and node-locked even if they are plain text files located on a remote file system. Only files that match a cryptographic hash of a registered creation may be exchanged, and for these the DRM enforces the security. Users may therefore create new documents or copy and paste from a registered creation to which they have access; it can only be forwarded to other users by registering the newly created file with the DRM. By restricting the set of permissible applications to those retaining file semantics in the transmission of data between systems (this is particularly the case for commonly used base protocols such as HTTP), this mechanism for tracking documents can also be extended to networked environments.

An extension of the COTS system's host network stack can analyze incoming and outgoing network streams and restrict data traffic to trusted hosts (for which the retention of semantic constraints is assumed) and therefore identify labels identifying creations that need to be traced. As a result, such encapsulated creations can be transmitted transparently without the need for changes in applications. Since the rights verification, tracking, and decryption of the data stream occur transparently within the operating system domain, application processes receiving such data streams can operate on simple plain text data streams.

If, however, the application (e.g., a World Wide Web (WWW) browser or similar WWW-enabled application) stores the datum again on a file system, the CIPRESS system can correlate the datum stored with the previously received data stream's label and thus automatically recreate the label for persistent storage. As a result, tracing and classification of documents extends beyond individual systems without requiring invasive modifications to either application programs or user behavior [23].

As noted before, no encryption system can protect against the possibility that a document is legitimately obtained and then converted to an analog representation only to be removed. As in the case of encrypting files and network traffic, it is irrelevant whether a legitimate copy falls into the hands of an adversary due to an oversight by a legitimate user or if the illegitimate removal of the analog copy is done deliberately. CIPRESS attempts to address this issue by embedding a digital watermarking mechanism in the operating system alongside the encryption mechanisms, which permits the backtracking of the creation's label by embedding all or part of the original creation's cryptographic hash value as the payload of the digital watermark. This mechanism ensures that for any registered creation for which a watermarking mechanism exists, a watermark identifying the user that retrieved the creation (i.e., the user for which the DRM has granted permission for a given operation and has generated the requisite audit information) is embedded into the document. This occurs regardless of the type of data access (i.e., from a file system or over a network connection) and takes place before the application, and hence the user has access to the creation. Any printout or even screen shot therefore contains the identity of the user; in addition to that, since the digital watermark used is capable of supporting hierarchical digital watermarks, the creation also contains two additional watermarks that are already embedded at the time of registration. One of these two server watermarks is a secret watermark known only to the administrator of the DRM system (typically this role belongs to an organizational security administrator).

The other watermark is the previously mentioned public watermark that can be read by anyone with the appropriate tool and allows the identification of the original (digital) document from the analog representation by extracting an identity label for the source database and a sufficiently large (48 bits) fragment of the cryptographic hash. This allows one to identify the digital source document even if only a fragment of a document (e.g., a cutout from a printed document) is available [24]. This watermarking mechanism is tightly integrated with both network and storage mechanisms outlined above and is schematically depicted in Figure 9.2.

Based on the labeling mechanism and the properties of the cryptographic hash algorithms used for identifying documents, this DRM architecture further permits not only the identification of creations but also the verification and protection of the integrity of creations by permitting the registration process to include the application of a digital signature to the original creation alongside its registration. As a result, the DRM database can be used to verify that a given datum (regardless of whether it is in plain text or in encapsulated form) matches a specific registered entity and the circumstances such as the identity and location of the user at the time of registration. In addition, the labeling mechanism also permits the retention of arbitrarily unique revisions by the DRM system, which represents the foundation for an efficient document management system that is also tightly integrated with the operating system.

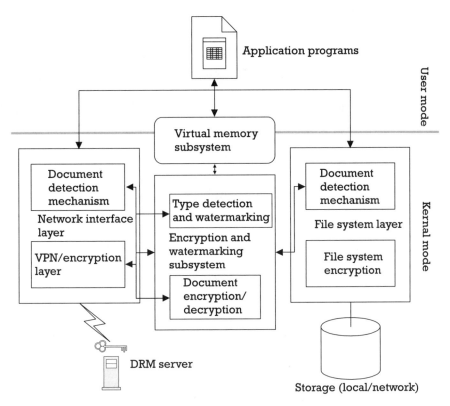

Figure 9.2   Component interaction for digital watermarking in CIPRESS.

9.1.4   Pragmatic aspects of digital rights management

Ostensibly, the purpose of digital rights management systems is ensuring that the rights of creators and other rights owners to multimedia content are protected in that technical means are used for the enforcement of copyright laws.

As partially made evident by the requirements or desiderata list in Section 9.1 and by the DRM systems both proposed and in existence, however, the goals for DRM-protected content are significantly more ambitious in that controls over the usage of the creations can be enforced that were not possible in the previous representations for such creations. Such fine-grained controls may be used to construct a number of different business and revenue generation models for rights owners. Examples of such models are the sale of time-restricted access to a creation (e.g., a movie rental that deactivates access to stored video data or the descrambling of the motion picture rental period has expired) or restrictions on the number of times that a creation can be used.

It can certainly be argued that application scenarios beneficial to both the customer and the rights owner can be created. One such scenario consists of providing trial versions that expire after a limited number of usages or providing only a sample segment of creation prior to payment. This can expose customers to new material they otherwise might not consider purchasing, thereby potentially increasing the revenue on the part of the rights owner (assuming, that is, the customer's budget for such creations or similar entertainment is not a priori fixed). On the other hand, customers can form a meaningful judgment over the desirability of creations prior to initiating purchases.

Similarly, traditional home video rentals relying on physical media can be replaced by an application scenario in which the motion picture is either downloaded under the protection of the DRM system and provided for the rental period or is provided via streaming media, the presentation again being under the protection of the DRM system.

The same underlying functionality can, however, be used to enhance the revenues of the rights owners while at the same time either restricting use of the protected material permitted in traditional representations bound to physical media (e.g., home video cassettes). In such a traditional setting, the number of times a motion picture is viewed during the rental period is beyond the control or knowledge of the entity providing the rented medium. With DRM, the possibility exists to restrict the rental period to a single usage or to require additional payment for multiple usage, as may be the case in situations where the original usage is interrupted or multiple usage is the norm rather than the exception (e.g., in the case of material intended for children).

Besides the potential price increases inherent in such scenarios, the DRM systems can also collect information that must be considered highly invasive of the privacy rights of individuals unless significant design and implementation efforts are expended to restrict such privacy invasions [25, 26]. This is of particular concern in environments where freedom of speech is impeded and where DRM systems can be abused for surveillance and similar oppressive measures. Furthermore, another revenue-enhancing scheme acting to the detriment of customers that has already been applied in the context of purchasing decisions and can be readily extended to information gleaned from DRM systems is price discrimination, that is, the variation of prices for the same product depending on the customer's purchasing habits and other information correlated to a given user, such as income levels.

The rental model has also been proposed for a number of media types and creations where, traditionally, technical and organizational reasons have precluded such. Examples specifically include digital libraries where access to creations is permitted to library patrons for a time-based fee [27–31], providing instant access to and superior searching and browsing capabilities in vast stores of knowledge.

Moreover, a number of commercial entities have initiated, for example, the provision of music on a subscription basis. Such services are intended to provide access to a large selection of creations for a flat fee. However, unlike most digital library systems in existence, the music subscription services model assumes that a DRM inhibits the storage and usage of the creations without a valid subscription. As an extension or alternative to such schemes, pay-per-use scenarios are also envisioned. Particularly the latter has the potential for placing customers at a disadvantage compared to the distribution of creations on physical media in that the amortized cost (unless there exists a threshold beyond which no further payments are required) for the customer is not bounded.

One of the most severe challenges for the acceptability of subscription or pay-per-use schemes is that of portability. Customers are unlikely to accept rights management schemes tied to specific devices that would require either separate purchase or licensing of creations or cumbersome transfer protocols (see Section 8.1), reducing not only the convenience customers are accustomed to but in some jurisdictions denying customers the right to create and use personal copies of creations once purchased or licensed.

As this "first sale" doctrine or its equivalent typically represents a fundamental barrier against the encroachment of copyright onto other property rights, it is also unlikely that legal frameworks will be changed to accommodate the desires of rights owners to the extent of permitting such far-reaching licensing conditions on the use of creations [32, 33]. This limitation, along with the possible loss of privacy due to the precise auditing and billing, particularly in pay-per-use scenarios, can be viewed as placing the customer at a significant disadvantage that may well balance or substantially outweigh any additional convenience for customers and, as a result, lead to an overall rejection of DRM-based business models and content.

Moreover, even if DRM systems are used for the enforcement of copyright regulations, the problem of different legal systems and hence of rights to a creation granted, for example, for personal use may differ significantly. This implies that a DRM system would need to accommodate the location that the user or device is, based on decisions as to what constitutes permissible operations or risk litigation as well as further loss of appreciation on the part of disadvantaged customers.

However, if a DRM system does support such nuanced control depending on locales, this in turn opens possible venues for attackers who can then select the most permissive locale or simulation thereof; similar scenarios were already exploited successfully in the context of locale-sensitive handling of encryption (the historical "you're in France" attack on encryption products that disabled all cryptographic functionality when operating in a French locale due to the prohibition of cryptography consisted simply of simulating a locale and thereby disabling cryptographic mechanisms).

The market for DRM systems appears to be sizable [34], with a large number of systems, primarily from industry, competing for this market—with most of the industry considering itself mainly as technology providers. Given the potential disadvantages for customers for some of the measures that can or are being taken based on DRM mechanisms discussed above, it is imperative for the users of DRM mechanisms that the models implemented strike a proper balance between the interests of customers and rights owners.

Unless such a balance is achieved—which appears particularly likely in the case of customers of multimedia content intended for personal use who have only limited individual leverage for contract negotiation—the creation of an adversarial relation between the contracting parties appears inevitable. However, given the inherent limitations of DRM mechanisms, particularly for the application scenario discussed in Section 9.2 and the fact that every DRM mechanism—even systems that were developed with considerable resources—so far has been broken [35], such use of DRM mechanisms may well be self-defeating.

9.2   Tamper protection

In most application scenarios discussed throughout this book, the medium, the device used for reproduction, or both, are physically available to a potential attacker. This implies that an attacker can observe any and all parts of a system's operation and modify the system and its interactions with the outside at will.

9.2.1   Techniques for software tamper protection

The prevention or detection of tampering with software has attracted research interest for some time [36] and has recently gained even more attention based on the use of byte-compiled programming languages and run-time environments such as Sun Microsystems' Java that lend themselves well to reverse engineering.

Despite the views espoused by Gosler [36] and the anecdotal evidence briefly outlined in Chapter 8, a number of patents for software protection mechanisms were filed and granted (e.g., [37–40]) that purported to afford software some measure of tamper resistance.

Loosely following the terminology from [41], the applicable protection techniques are, on one hand, obfuscation techniques which transform a given program into a semantically equivalent function which is harder to reverse-engineer, and on the other hand tamper-detection techniques[2] striving to cause a program to malfunction or cease operations altogether if and when it detects that it has been modified. These techniques are complementary in that the tamper detection and any reaction mechanism related to it should also be protected from reverse-engineering itself that could lead to the deactivation of such code.

Automated obfuscation techniques can occur at the level of control structures [42] in creating predicates that result in constructs that are difficult to analyze statically, or they can occur at the data structure level [43] (e.g., by transforming vector records into arbitrary permutations or splitting variables into additive components and vice versa). However, Barak et al. have shown the sharp limitations on any obfuscation technique and have obtained several impossibility results that not only apply to programs but also to circuits [44], showing that there exist classes of programs that cannot be obfuscated. The proponents of obfuscation, however, do not provide a rigorous analysis of the work factor to be required of an adversary or consider the possibility of inverting or partially inverting the obfuscation process under the assumption that an adversary has knowledge of the obfuscation mechanism used.

Tamper detection in software can occur by verifying constraints on intermediate computation results [45, 46]; early approaches, however, resulted in multiplicative factors as computational overhead [47]. Later probabilistic approaches to certain subtypes of computations achieve sublinear computational overhead [48]. While the suitability of such consistency checks represents a valuable tool for program validation, its suitability as a tamper detection mechanism is limited by the observability of the verification steps, since an adversary may alter not only the program itself but also any conditions imposed by altering the control flow or data representation to match a successful validation step.

Another technique for tamper resistance was proposed by Aucsmith [40, 49]. It proposes the construction of integrity verification kernels (IVK), which can operate independently or act as multiple mutually verifying instances. The mechanisms employed by the IVK are:

---

2. This is referred to as *tamper-proofing* by [41], which does appear to be a somewhat unfortunate choice.

> *Interleaving*   IVK calculations are separated into multiple subtasks such that verification steps are executed concurrently to avoid situations where an attacker can replace the result of a verification step with a precalculated result.

> *External secret*   Calculations of the IVK occur based on a secret known only to the IVK itself.

> *Obfuscation*   IVK are self-modifying and self-encrypting, reusing memory for different variables at different times.

> *Instance specificity*   Each IVK instance contains data specific to the particular instance that affects the behavior of the IVK.

> *Nondeterminism*   Multithreading is used to introduce an element of nondeterminism.

The mechanism described above incorporates mechanisms applied unsuccessfully two decades earlier in copy protection; from the perspective of an attacker controlling the software component and system (against which the mechanism is a priori not adequate, as Aucsmith notes), such an IVK has the additional advantage of providing a well-defined segment of code that can be analyzed and then circumvented.

9.2.2 Techniques for hardware tamper protection

Software-based techniques for tamper resistance are, as discussed in Section 9.2.1, inadequate for protection against attacks where the adversary has complete control over the physical and software environment of the content and, moreover, the protection system itself.

Denying the adversary the control over the content and protection system in the form of physical security is therefore a self-evident prerequisite for the accomplishment of the overall protection objective. In situations where devices or subsystems are by definition under the control of an adversary, this implies that additional technical means are required to deny access [50, 51]. Physical security or tamper protection can be subdivided into the following three objectives.

1. *Tamper resistance*   Resistance constitutes the basic mechanism for any tamper protection. The adversary is denied access to or the ability to physically manipulate the protected system. The system or subsystem to be protected must be brought within a well-defined defensive

perimeter or enclosure where any unauthorized access or manipulation can be accomplished only by the expenditure of a considerable work factor over an unprotected equivalent.

2. *Tamper response* Assuming the existence of a tamper-resistant system and further assuming that an adversary has the capability and desire to break the tamper resistance mechanism, tamper-responding mechanisms are required if the application scenario calls for maintaining the integrity of a system (or the disabling in case the integrity is violated). While in most situations responses such as alarms are possible, this is typically not an option for the application scenarios considered here, as devices under the control of an adversary will not be able to transmit alerts. Instead, responses must center on disabling data and functionality that can be used to extract content data or reverse-engineer the mechanisms required for the extraction of content data such as key material [52].

3. *Tamper evidence* Continued trust in a system or component that was compromised and modified to perform operations can lead to significant damage as data (content data, key, and authentication materials for cryptographic processing) can continue to be compromised. Physical protection mechanisms include frangible seals surrounding the physical enclosure or chemical protection mechanisms (e.g., ink held between impermeable layers that leaks once a surrounding layer has been violated). As with alerts in the case of tamper response mechanisms, the applicability to typical content protection scenarios is limited, since tamper evidence is relevant only if an audit mechanism exists that permits the inspection of possible evidence of tampering and subsequent actions (e.g., barring a device from further communication by entering it into a blacklist).

From the above description, a physical protection mechanism must provide both tamper resistance and tamper responsive behavior; a constraint on any such technique is typically that the cost for providing a large enclosure with tamper resistance is prohibitive.

Tamper resistance in hardware has been the subject of classified research since the early 1960s when, in accordance with NSAM 160 [53], U.S. nuclear weapons were to be prepositioned within Europe and under the control of NATO allies. Devices were necessary that provided inhibitory control over the firing of the warheads to ensure compliance with U.S. law and to centralize control over land- and air-based weapons that would previously have rapidly devolved to the squad level in tactical situations [54].

These devices, called PAL (permissive action links; originally, prescribed action link [54, 55]) were developed at Sandia National Laboratories first in the form of simple electromechanical combination locks (CAT A) and later evolved to sophisticated electronic devices permitting the unlocking of multiple weapons using a single transmission (CAT F). The PALs are tamper-resistant devices that are tightly integrated with the actual weapons system [56] and contain elaborate self-destruct mechanisms that incapacitate the firing mechanism and also destruct the PAL mechanism itself to prevent reverse engineering if tampering is detected [57, 58]. The desired complexity for an adversary can be postulated as "Bypassing a PAL should be, as one weapons designer graphically put it, about as complex as performing a tonsillectomy while entering the patient from the wrong end" [59].

Not only is violent self-destruction hardly an option for content protection, the cost of the protection mechanism in relation to the risk averted is also markedly different. Also, the acceptability of accidental triggering of tamper-resistant mechanisms is markedly different in consumer devices, where such an event will be seen as a product defect. Such a perceived defect will result in costs for replacement and handling and may also be to the detriment of the manufacturer's reputation. Such considerations sharply reduce the capabilities available to designers for tamper protection.

9.2.2.1 Possible attacks

Breaking through the protected physical perimeter can be accomplished at several levels. At the component or subsystem level, this typically involves various machining methods that attempt to remove inert protective mechanisms without causing sensors to register and subsequently initiate tamper-responding behavior. These attacks are used as a prelude to a subsequent attack that probes or manipulates the circuit or subsystem; these attacks are described below.

Machining can occur with a variety of tools depending on the type of enclosure (e.g., a simple steel shell or a resinous potting material) and the feature size at which sensors and defensive mechanisms need to be circumvented. In the simplest case, the careful manual use of surgical scalpels and drills may suffice for opening larger subsystem enclosures, although micromanipulators commonly used in biomedical research can be used for smaller enclosures; this, however, may have the disadvantage of using conducting materials that may cause defensive mechanisms to react. High-pressure water jets can also be used for micromachining in such cases. If pressure or vibration sensors would be triggered by such machining, ablation using suitable laser devices provides another alternative.

The removal of enclosures, particularly of integrated circuits, is a common procedure for quality control and is frequently accomplished by etching methods that use solvents or acids that react with the enclosure (e.g., potting material) but not with the targeted circuit (or the passivation layers on the surface of the circuit) itself. The solvent must be applied carefully and in a steady stream (e.g., under pressure) to remove reaction byproducts from the targeted surface. Examples of such solvents include HNO_3 or supercritical fluid CO_2, although etching of live circuits presents problems due to the conductivity of the etching agent. For circuits protected by etching-resistant components, ion beam workstations cannot only be used for high-resolution surface analysis, but also for the targeted kinetic removal of such layers.

Another type of attack that can be employed is the manipulation of environmental characteristics to induce erroneous behavior such as timing faults, gates not operating properly, or disabling the overwrite circuitry for non-volatile memory. One of the simplest such attacks is the targeted variation of the supply voltages (high and low) which can also be made time-variable to induce certain desirable behavior such as a misinterpretation of instructions. Due to slight variations in the characteristics of individual junctions and transistors, such faults can be induced selectively and in a reproducible fashion. Besides the supply voltages, other signals can also be used to induce faults; one of the most commonly used techniques is the introduction of clock glitches by shortening or lengthening clock intervals and modifying signal edges; this can, for example, be accomplished by using a signal generator in conjunction with an in-circuit emulator. Other, cruder environmental attacks include changes in operating temperature or the deliberate introduction of delays to external interfaces. Rapid temperature changes (i.e., freezing using supercooled liquid gases with desirable heat transfer characteristics) can also help to both preserve the contents of memory that would be erased by tamper response measures and disable other circuitry in the process due to latency characteristics of, for example, commonly used CMOS RAM circuits. Similar imprinting can also be accomplished by exposing the relevant circuit elements to targeted radiation (x-ray bands depending on the feature size of the circuit). While particularly the latter will result in the disabling of the device, this may be acceptable if only access to secrets such as decryption keys is desired.

Given a breached enclosure, the recovery of secret information stored in the enclosure (e.g., cipher keys, authentication data) or reverse engineering of the circuit can be accomplished by probing techniques of varying levels of intrusiveness [60]. In simple cases, a passive probe may consist merely of a logic analyzer attached to various signals of the circuit. Commercial equipment for test and quality control of integrated circuits is readily available and can be used to conduct such analysis via microprobes directly on the surface of the circuit. Such probes can also be used for injecting voltage levels at predefined locations and times into a circuit, permitting at the very least the introduction of faults into the circuit, although direct circumvention of protective circuitry is also possible if the design of the protection mechanism is well known. Reverse-engineering techniques both for the recovery of secret data and duplication of circuits include optical microscopy and interferometry electron beam backscattering, scanning and transmission electron microscopy, or even atomic force microscopy [61]. The presence of protective layers impermeable to laser or electron backscattering analysis can be overcome by systematically removing layers from the circuit using plasma etching. The actual process of generating circuit diagrams (particularly those of ROM contents) from the images thus obtained can be largely automated; a number of commercial service providers offer the complete reverse engineering of integrated circuits. Scanning electron microscopes and laser scanners can, in addition, be manipulated in such a way that arbitrary signals can be both read and written on a given circuit; this can to some extent even be accomplished without depassivation of the desired features and connects required for other probing attacks.

An important class of eavesdropping attack is power analysis. These techniques make use of the fact that the CMOS circuitry used almost exclusively in modern systems has power consumption characteristics that directly relate to operations performed [62–71]. Besides power consumption, the timing particularly of cryptographic operations, such as those used in conditional access mechanisms, also permits conclusions about the key material used without invasive techniques [72].

A number of these attacks can be performed using very limited resources or at most using tools readily available in many university research facilities [73, 74]; it must also be noted that the interfaces of such a tamper-resistant device may themselves become the target of attacks, permitting types of attacks through software that would otherwise require the circumvention of tamper-resistance mechanisms [75].

9.2.2.2 Defensive mechanisms

Defensive mechanisms can again be grouped into the three categories of tamper resistance, response, and evidence, although in some cases the application of a defensive mechanism may serve more than one goal.

Immediate tamper resistance is achieved through enclosures such as metal or ceramic shieldings. Such protection is mainly effective if tamper response mechanisms are also available, since any physical barrier can be breached given adequate time and resources such as drills or acids. The size, weight, and cost of materials typically limits the extent of such protection mechanisms. Cost is also an implicit factor in that thus protected elements are inherently nonserviceable, providing an additional incentive to keep the size of tamper-resistant enclosures as small as possible. However, this size limitation implies that critical functionality is frequently left outside of the enclosure, permitting attackers other venues of approach, such as the interfaces of the enclosure or locations where the desired data is available in plain text.

Similar approaches can also be pursued at the circuit level by adding resist layers, routing various undesirable traces at the outside of a circuit, or covering the circuit with a chemical coating that prevents probing and would require chemicals for removal that destroy the functional components at the same time. Another approach based on materials that can counter analysis and manipulation via laser scanners is the use of substrate materials that are not transparent to such scanners (such as Silicon-on-Sapphire). Particularly the latter mechanism raises the cost for both defender and attacker considerably.

As noted above, tamper resistance mechanisms can only be employed effectively if a corresponding tamper-responding mechanism is also part of the defensive system. Such sensors can monitor the environmental conditions of the enclosure to determine if an attack occurs; examples of parameters include acceleration, motion, and vibration sensors that respond to drilling or other crude attempts at breaking the enclosure, temperature sensors that determine whether a module is taken outside its operating range (e.g., to induce glitches or to use memory remanence effects on cold circuits), photoreceptors and pressure sensors accompanied by an enclosure with a pressure differential to the outside determining the breaching of the enclosure, and various other radiation sensors that can detect attempts at probing or imprinting via radiation.

Another approach to ascertain the integrity of the enclosure is the active use of enclosure materials as sensors. By integrating an extremely thin wire into the perimeter of the enclosure (e.g., in a resinous potting material) in such a way that probes cannot be inserted without dislocating or severing a wire, monitoring of the resistance of the wire provides an indication of the integrity of the enclosure itself. A more advanced variant of the approach consists of using not wires but modification (e.g., by chemical altering) of the potting material or a highly flexible substrate such that it is conducting, thereby reducing the contrast of the protective conductors.

This mechanism can also be employed at the circuit level, although the wire or wires will then be arranged in a largely two-dimensional pattern—making the location easier to predict and to circumvent—since the dispersion in depth by potting material for larger enclosures is not available.

Similarly, various enclosure materials including glass or plastics can be used as sensors by measuring the piezoelectric effect of stresses exerted on the enclosure. Since this effect requires a minimum kinetic force to be measurable, it is highly desirable to use a material such as prestressed (tempered) glass.

Moreover, glass or similar transparent enclosure materials can also be used as indirect sensors in analyzing scattering effects due to stresses, fracture, or dislocation from well-defined light sources. Another highly sensitive and therefore also problematic approach due to false positive alarms is the use of interferometric sensors based on enclosures; as in the case of scattering, a significant deviation from a preestablished norm is taken as an attack. Such deviations can, however, also be caused by temperature variations or minor vibrations due to, for example, fans within the vicinity of the tamper-resistant device. At a lower level of sensitivity and hence rate of false positives, the detection of deflection angles of mirroring surfaces (nominally incident at a well-defined angle) can also be used.

To protect against the deliberate induction of glitches, attempts at inducing remanence, or causing on-circuit fuses to be activated, voltage sensors can be deployed at the enclosure and also at the circuit level to monitor any changes from the accepted operating envelope for all components. This, however, is particularly error prone in situations where battery or even mains power is not reliable or well defined.

Physical destruction as a means of tamper response will, unlike the case of PAL components, rarely be an option, although this does represent the only known reliable technique for erasing sensitive memory contents. Such destruction must occur very quickly to prevent the employment of countermeasures (e.g., in the case of chemical dissolution of circuits or heating). This limits the destructive means to approaches such as explosives or electric discharges that are difficult and dangerous to handle and may not be permitted in devices for civilian use; however, even in case explosives are used, it is important to ensure a minimum fragment size of the destroyed circuitry to ensure that neither reverse engineering nor data remanence can be exploited by adversaries.

More benign response mechanisms include the erasing (typically through multiple overwrites) of critical memory areas or—depending on the memory circuit type used—separating memory areas from power supplies. Somewhat problematic, particularly in the latter case, is the issue of data remanence, particularly at lower temperatures; simple elimination of the power supply is therefore rarely sufficient.

Tamper evidence mechanisms are, as noted above, not typically applicable for the scenarios considered here; possible tamper-evident mechanisms include the use of brittle or frangible materials (e.g., ceramic) that cannot be reconstituted (e.g., through melting within parameters covered by the operating range of the protected device). Another option is the use of unique surface or material structures that cannot be reproduced or recovered easily, such as highly polished machined surfaces. In all such cases, the problem of benign, accidental damage to the tamper-protected enclosure is difficult to deal with in typical commercial environments, where the compound cost of replacement or deactivation of individual devices is considerable compared to the risk of disclosure.

Besides such passive surfaces, another tamper-evident mechanism that can also be coupled with tamper-responding features is the use of *bleeding paint*, that is, sheets of material either enclosing a bulk liquid or a liquid encapsulated in microscopic bubbles that permeate the surrounding area if violated. If such a liquid or paint conducts electricity, it can also act as a probe detector, although both can work only in conjunction with temperature sensors, since otherwise local freezing around the site of a probe can prevent the dissipation of the liquid.

In the case of power or timing analysis, the protection mechanisms required go significantly beyond the protecting enclosure, in this case, an obfuscation strategy must be used that takes into consideration aspects such as the energetic efficiency of algorithms based on the parameters used and may also require the use of algorithms markedly different from naive implementations [76–81].

References

[1] International DOI Foundation, *The DOI Handbook*, version 2.3.0, August 2002.

[2] infoMarket Business Development Group, *Cryptolope Containers*, technical report, International Business Machines Corporation, 1997.

[3] ContentGuard, *eXtensible rights Markup Language (XrML) Core 2.1 Specification*, White Paper, ContentGuard Holdings, 2002.

[4] Pearson, W., "Digital Transmission Content Protection," oral presentation, June 1999.

[5]   Hitachi, Ltd., et al., *Digital Transmission Content Protection Specification*, version 1.2, July 2001 (informational version).

[6]   IEEE, *IEEE STD 1363-2000: Standard Specifications for Public-Key Cryptography*, January 2000.

[7]   NIST, *Digital Signature Standard*, National Institute of Standards and Technology, Gaithersburg, MD, May 1994.

[8]   NIST, *Data Encryption Standard (DES)*, National Institute of Standards and Technology, Gaithersburg, MD, October 1999. Supersedes FIPS PUB 46-2 (1993).

[9]   Electronic Frontier Foundation, *Cracking DES: Secrets of Encryption Research, Wiretap Politics & Chip Design*, Sebastopol, CA: O'Reilly & Associates, 1998.

[10] Kelsey, J., B. Schneier, and D. Wagner, "Mod *n* Cryptanalysis, with Applications Against RC5P and M6," in L. Knudsen, (ed.), *Fast Software Encryption: 6th International Workshop (FSE '99)*, Springer Verlag: Rome, March 1999, pp. 139–155.

[11] Intel Corporation, *High-Bandwidth Digital Content Protection System*, version 1.0, February 2000 (informational version).

[12] Crosby, S., et al., "A Cryptanalysis of the High-Bandwidth Digital Content Protection System," in T. Sander, (ed.), *Security and Privacy in Digital Rights Management: ACM CCS-8 Workshop DRM 2001*, Vol. 2320, Heidelberg, Germany: Springer Verlag, 2002, pp. 192–200.

[13] Compaq Computer Corporation, et al., *Trusted Computing Platform Alliance (TCPA) Main Specification*, version 1.1b, February 2002.

[14] Arbaugh, W., "Improving the TCPA Specification," *IEEE Computer*, Vol. 35, No. 8, August 2002, pp. 77–79.

[15] Anderson, R., "Security in Open Versus Closed Systems—The Dance of Boltzmann, Coase and Moore," *Proceedings of Open Source Software: Economics, Law and Policy*, Toulouse, France, June 2002.

[16] Compaq Computer Corporation, et al., *TCPA Specification/TPM Q & A*, July 2002.

[17] ISO/IEC Joint Technical Committee 1 Subcommittee 27 Working Group 3: Security Evaluation Criteria, *Information Technology—Security Techniques—Evaluation Criteria for IT Security—Part 1: Introduction and General Model*, ISO/IEC 15408-1, December 1999, version 2.1.

[18] ISO/IEC Joint Technical Committee 1 Subcommittee 27 Working Group 3: Security Evaluation Criteria, *Information Technology—Security Techniques—Evaluation Criteria for IT security—Part 2: Security Functional Requirements*, ISO/IEC 15408-2, December 1999, version 2.1.

[19] ISO/IEC Joint Technical Committee 1 Subcommittee 27 Working Group 3: Security Evaluation Criteria, *Information Technology—Security Techniques—Evaluation Criteria for IT Security—Part 3: Security Functional Requirements*, ISO/IEC 15408-3, December 1999, version 2.1.

[20] Reynolds, F., and J. Heller, "Kernel Support for Network Protocol Servers," *Proceedings of the USENIX Mach Symposium*, Monterey, CA, November 1991, pp. 149–162.

[21] Jones, M. B., "Interposition Agents: Transparently Interposing User Code at the System Interface," *ACM Operating Systems Review*, Vol. 27, No. 5, December 1993, pp. 80–93.

[22] Busch, C., et al., "A System for Intellectual Property Protection," *Proceedings of the World Multiconference on Systemics, Cybernetics, and Informatics (SCI 2000)/International Conference on Information Systems Analysis and Synthesis (ISAS 2000)*, Orlando, FL, July 2000, pp. 225–230.

[23] Rademer, E., and S. Wolthusen, "Transparent Access to Encrypted Data Using Operating System Network Stack Extensions," in R. Steinmetz, J. Dittman, and M. Steinebach, (eds.), *Communications and Multimedia Security Issues of the New Century: Proceedings of the IFIP TC6/TC11 Fifth Joint Working Conference on Communications and Multimedia Security (CMS'01)*, Darmstadt, Germany, May 2001, pp. 213–226.

[24] Busch, C., and S. Wolthusen, "Tracing Data Diffusion in Industrial Research with Robust Watermarking," in J.-L. Dugelay and K. Rose, (eds.), *Proceedings of the 2001 Fourth Workshop on Multimedia Signal Processing (MMSP'01)*, Cannes, October 2001, pp. 207–212.

[25] Davis, R., *The Digital Dilemma: Intellectual Property in the Information Age*, Washington D.C.: National Academy Press, 2000.

[26] Feigenbaum, J., et al., "Privacy Engineering for Digital Rights Management Systems," in T. Sander, (ed.), *Security and Privacy in Digital Rights Management: ACM CCS-8 Workshop DRM 2001*, Vol. 2320 of *Lecture Notes in Computer Science*, Philadelphia: Springer-Verlag, 2002, pp. 76–105.

[27] Samuelson, P., "Encoding the Law into Digital Libraries," *Communications of the Association for Computing Machinery*, Vol. 41, No. 4, 1998, pp. 13–18.

[28] Levy, D. M., and C. C. Marshall, "Going Digital: A Look at Assumptions Underlying Digital Libraries," *Communications of the Association for Computing Machinery*, Vol. 38, No. 4, April 1995, pp. 77–84.

[29] Wiederhold, G., "Digital Libraries, Value, and Productivity," *Communications of the Association for Computing Machinery*, Vol. 38, No. 4, April 1995, pp. 85–96.

[30] Adam, N., and Y. Yesha, "Strategic Directions in Electronic Commerce and Digital Libraries: Towards a Digital Agora," *ACM Computing Surveys*, Vol. 28, No. 4, December 1996, pp. 818–835.

[31] Sistla, A. P., et al., "Towards a Theory of Cost Management for Digital Libraries and Electronic Commerce," *ACM Transactions on Database Systems*, Vol. 23, No. 4, December 1998, pp. 411–452.

[32] Black, J., "A Bad, Sad Hollywood Ending?" *BusinessWeek*, May 2002.

[33] Salkever, A., "Guard Copyrights, Don't Jail Innovation," *BusinessWeek*, March 2002.

[34] Stamp, M., "The Risks of Digital Rights Management," *Communications of the Association for Computing Machinery*, Vol. 45, No. 9, September 2002, p. 120.

[35] Huang, A., "Keeping Secrets in Hardware: The Microsoft XBox Case Study," technical report, AI Memo 2002-008, Massachusetts Institute of Technology Artificial Intelligence Lab, Cambridge, May 2002.

[36] Gosler, J. R., "Software Protection: Myth or Reality?" in Williams, H. C., (ed.), *Advances in Cryptology (CRYPTO '85)*, Vol. 218 of *Lecture Notes in Computer Science*, Santa Barbara, CA: Springer-Verlag, August 1985, pp. 140–157.

[37] Holmes, K., "Computer Software Protection," United States Patent 5,287,407, May 1991. Granted in February 1994.

[38] Davidson, R. I., and N. Myhrvold. "Method and System for Generating and Auditing a Signature for a Computer Program," United States Patent 5,745,569, June 1994. Granted in September 1996.

[39] Moskowitz, S. A., and M. Cooperman, "Method for Stega-Cipher Protection of Computer Code," United States Patent 5,745,569, January 1996. Granted in April 1998.

[40] Aucsmith, D., and G. Graunke, "Tamper Resistant Methods and Apparatus," United States Patent 5,892,899, June 1996. Granted in April 1999.

[41] Collberg, C., and C. Thomborson, "Watermarking, Tamper-Proofing, and Obfuscation—Tools for Software Protection," Technical Report 2000-03, Department of Computer Science, University of Arizona, Tucson, 2000. Also, University of Auckland Computer Science Technical Report #170.

[42] Collberg, C. S., C. D. Thomborson, and D. Low, "Manufacturing Cheap, Resilient, and Stealthy Opaque Constructs," *Proceedings of the 25th ACM SIGPLAN-SIGACT Symposium on Principles of Programming Languages (POPL '98)*, San Diego, CA, 1998, pp. 184–196.

[43] Collberg, C. S., C. D. Thomborson, and D. Low, "Breaking Abstractions and Unstructuring Data Structures," *Proceedings of the 1998 International Conference on Computer Languages (ICCL '98)*, Chicago, May 1998, pp. 28–38.

[44] Barak, B., et al., "On the (Im)possibility of Obfuscating Programs," in J. Kilian, (ed.), *Advances in Cryptology, (CRYPTO 2001)*, Vol. 2139, Santa Barbara, CA: Springer Verlag, 2001, pp. 1–18.

[45] Blum, M., "Program Checking," in S. Biswas and K. V. Nori, (eds.), *Proceedings of Foundations of Software Technology and Theoretical Computer Science, 11th Conference*, Vol. 560 of *Lecture Notes in Computer Science*, New Delhi, India: Springer-Verlag, December 1991, pp. 1–9.

[46] Blum, M., "Program Result Checking: A New Approach to Making Programs More Reliable," in A. Lingas, R. G. Karlsson, and S. Carlsson, (eds.), *Proceedings of*

Automata, Languages and Programming, 20nd International Colloquium (ICALP '93), Vol. 700 of *Lecture Notes in Computer Science*, Lund, Sweden: Springer-Verlag, July 1993, pp. 1–14.

[47] Blum, M., M. Luby, and R. Rubinfeld, "Self-Testing/Correcting with Applications to Numerical Problems," *Journal of Computer and System Sciences*, Vol. 47, No. 3, December 1993, pp. 549–595.

[48] Ergün, F., et al., "Spot-Checkers," *Proceedings of the Thirtieth Annual ACM Symposium on the Theory of Computing*, Dallas, May 1998, pp. 259–268.

[49] Aucsmith, D., "Tamper Resistant Software: An Implementation," in R. J. Anderson, (ed.), *Information Hiding: First International Workshop*, Vol. 1174 of *Lecture Notes in Computer Science*, Cambridge, U.K.: Springer-Verlag, 1996, pp. 317–333.

[50] Clark, A. J., "Physical Protection of Cryptographic Devices," in D. Chaum and W. L. Price, (eds.), *Advances in Cryptology (EUROCRYPT '87)*, Vol. 304 of *Lecture Notes in Computer Science*, Amsterdam: Springer-Verlag, April 1988, pp. 83–93.

[51] Weingart, S. H., "Physical Security Devices for Computer Subsystems: A Survey of Attacks and Defenses," in Ç. Koç and C. Paar, (eds.), *Cryptographic Hardware and Embedded Systems—CHES 2000, Proceedings of the Second International Workshop*, Vol. 1965 of *Lecture Notes in Computer Science*, Worcester, MA: Springer-Verlag, August 2000, pp. 302–317.

[52] Chaum, D., "Design Concepts for Tamper Responding Systems," in D. Chaum, (ed.), *Advances in Cryptology (CRYPTO '83)*, Santa Barbara, CA, August 1983, pp. 387–392.

[53] Kennedy, J. F., *Permissive Links for Nuclear Weapons in NATO*, NSAM 160, Washington D.C., 1962.

[54] Stein, P., and P. D. Feaver, *Assuring Control of Nuclear Weapons*, Technical Report CSIA Occasional Paper No. 2, Center for Science and International Affairs, Harvard University, Cambridge, MA, 1987.

[55] Sandia National Laboratories, *PAL Control of Theater Nuclear Weapons*, Technical report SAND82-2436, Sandia National Laboratories, Albuquerque, NM, 1982.

[56] Sandia National Laboratories, *Survey of Weapon Development and Technology*, Technical Report WR-708, Sandia National Laboratories, Albuquerque, NM, 1985.

[57] Caldwell, D., "Permissive Action Links," *Survival*, Vol. 29, No. 3, May 1987, pp. 224–238.

[58] Cotter, D. R., "Peacetime Operations: Safety and Security," in A. B. Carter, J. D. Steinbruner, and C. A. Zraket, (eds.), *Managing Nuclear Operations*, Washington D.C.: Brookings Institution Press, 1987, pp. 17–54.

[59] Caldwell, D., and P. D. Zimmerman, "Reducing the Risk of Nuclear War with Permissive Action Links," in B. M. Blechman, (ed.), *Technology and the*

Limitation of International Conflict, Lanham, MD: University Press of America, 1989, pp. 137–150.

[60] Handschuh, H., P. Paillier, and J. Stern, "Probing Attacks on Tamper-Resistant Devices," in Ç. Koç and C. Paar, (eds.), *Cryptographic Hardware and Embedded Systems—CHES 1999. Proceedings of the First International Workshop*, Vol. 1717 of *Lecture Notes in Computer Science*, Worcester, MA: Springer-Verlag, August 1999, pp. 303–315.

[61] Avery, L. R., et al., "Reverse Engineering Complex Application-Specific Integrated Circuits (ASICs)," *Proceedings of Diminishing Manufacturing Sources and Material Shortages Conference (DMSMS 2002)*, New Orleans, March 2002.

[62] Kocher, P., J. Jaffe, and B. Jun, "Differential Power Analysis," in M. J. Wiener, ed., *Advances in Cryptology (CRYPTO '99)*, Vol. 1666 of *Lecture Notes in Computer Science*, Santa Barbara, CA: Springer-Verlag, August 1999, pp. 388–397.

[63] Messerges, T. S., E. A. Dabbish, and R. H. Sloan, "Power Analysis Attacks of Modular Exponentiation in Smartcards," in Ç. Koç and C. Paar, (eds.), *Cryptographic Hardware and Embedded Systems—CHES 1999. Proceedings of the First International Workshop*, Vol. 1717 of *Lecture Notes in Computer Science*, Worcester, MA: Springer-Verlag, August 1999, pp. 144–157.

[64] Goubin, L., and J. Patarin, "DES and Differential Power Analysis (The "Duplication" Method)," in Ç. Koç and C. Paar, (eds.), *Cryptographic Hardware and Embedded Systems—CHES 1999. Proceedings of the First International Workshop*, Vol. 1717 of *Lecture Notes in Computer Science*, Worcester, MA: Springer-Verlag, August 1999, pp. 158–172.

[65] Fahn, P. N., and P. K. Pearson, "IPA: A New Class of Power Attacks," in Ç. Koç and C. Paar, (eds.), *Cryptographic Hardware and Embedded Systems—CHES 1999. Proceedings of the First International Workshop*, Vol. 1717 of *Lecture Notes in Computer Science*, Worcester, MA: Springer-Verlag, August 1999, pp. 173–186.

[66] Mayer-Sommer, R., "Smartly Analyzing the Simplicity and the Power of Simple Power Analysis on Smartcards," in Ç. Koç and C. Paar, (eds.), *Cryptographic Hardware and Embedded Systems—CHES 2000. Proceedings of the Second International Workshop*, Vol. 1965 of *Lecture Notes in Computer Science*, Worcester, MA: Springer-Verlag, August 2000, pp. 78–92.

[67] Anwarul Hasan, M., "Power Analysis Attacks and Algorithmic Approaches to Their Countermeasures for Koblitz Curve Cryptosystems," in Ç. Koç and C. Paar, (eds.), *Cryptographic Hardware and Embedded Systems—CHES 2000. Proceedings of the Second International Workshop*, Vol. 1965 of *Lecture Notes in Computer Science*, Worcester, MA: Springer-Verlag, August 2000, pp. 93–108.

[68] Schindler, W., "A Timing Attack Against RSA with the Chinese Remainder Theorem," in Ç. Koç and C. Paar, (eds.), *Cryptographic Hardware and Embedded Systems—CHES 2000. Proceedings of the Second International Workshop*, Vol. 1965 of *Lecture Notes in Computer Science*, Worcester, MA: Springer-Verlag, August 2000, pp. 109–124.

[69] Messerges, T. S., "Using Second-Order Power Analysis to Attack DPA Resistant Software," in Ç. Koç and C. Paar, (eds.), *Cryptographic Hardware and Embedded Systems—CHES 2000, Proceedings of the Second International Workshop*, Vol. 1965 of *Lecture Notes in Computer Science*, Worcester, MA: Springer-Verlag, August 2000, pp. 238–251.

[70] Clavier, C., J.-S. Coron, and N. Dabbous, "Differential Power Analysis in the Presence of Hardware Countermeasures," in Ç. Koç and C. Paar, (eds.), *Cryptographic Hardware and Embedded Systems—CHES 2000. Proceedings of the Second International Workshop*, Vol. 1965 of *Lecture Notes in Computer Science*, Worcester, MA: Springer-Verlag, August 2000, pp. 252–261.

[71] Schindler, W., "A Combined Timing and Power Attack," in D. Naccache and P. Paillier, (eds.), *Proceedings of Public Key Cryptography, 5th International Workshop on Practice and Theory in Public Key Cryptosystems (PKC 2002)*, Vol. 2274 of *Lecture Notes in Computer Science*, Paris: Springer-Verlag, February 2002, pp. 263–279.

[72] Kocher, P. C., "Timing Attacks on Implementations of Diffie-Hellman, RSA, DSS, and Other Systems," in N. Koblitz, (ed.), *Advances in Cryptology (CRYPTO '96)*, Vol. 1109 of *Lecture Notes in Computer Science*, Santa Barbara, CA: Springer-Verlag, August 1996, pp. 104–113.

[73] Anderson, R., and M. Kuhn, "Tamper Resistance—a Cautionary Note," *Proceedings of the 2nd USENIX Workshop on Electronic Commerce*, Oakland, CA, November 1996, pp. 1–11.

[74] Anderson, R. J., and M. G. Kuhn, "Low Cost Attacks on Tamper Resistant Devices," in B. Christianson et al., (eds.), *Proceedings of the 5th Security Protocols International Workshop*, Vol. 1361 of *Lecture Notes in Computer Science*, Paris: Springer-Verlag, April 1997, pp. 125–136.

[75] Bond, M., "Attacks on Cryptoprocessor Transaction Sets," in Ç. Koç, D. Naccache, and C. Paar, (eds.), *Cryptographic Hardware and Embedded Systems—CHES 2001. Proceedings of the Third International Workshop*, Vol. 2162 of *Lecture Notes in Computer Science*, Paris: Springer-Verlag, May 2001, pp. 220–234.

[76] von Willich, M., "A Technique with an Information-Theoretic Basis for Protecting Secret Data from Differential Power Attacks," in B. Honary, (ed.), *Proceedings of Cryptography and Coding, 8th IMA International Conference*, Cirencester, U.K., December 2001, pp. 44–62.

[77] Oswald, E., and M. Aigner, "Randomized Addition-Subtraction Chains as a Countermeasure Against Power Attacks," in Ç. Koç, D. Naccache, and C. Paar, (eds.), *Cryptographic Hardware and Embedded Systems—CHES 2001. Proceedings of the Third International Workshop*, Vol. 2162 of *Lecture Notes in Computer Science*, Paris: Springer-Verlag, May 2001, pp. 39–50.

[78] May, D., H. L. Muller, and N. P. Smart, "Random Register Renaming to Foil DPA," in Ç. Koç, D. Naccache, and C. Paar, (eds.), *Cryptographic Hardware and Embedded Systems—CHES 2001. Proceedings of the Third International Workshop*,

Vol. 2162 of *Lecture Notes in Computer Science*, Paris: Springer-Verlag, May 2001, pp. 28–38.

[79] Coron, J.-S., "Resistance Against Differential Power Analysis for Elliptic Curve Cryptosystems," in Ç. Koç and C. Paar, (eds.), *Cryptographic Hardware and Embedded Systems—CHES 1999. Proceedings of the First International Workshop*, Vol. 1717 of *Lecture Notes in Computer Science*, Worcester, MA: Springer-Verlag, August 1999, pp. 292–302.

[80] Coron, J.-S., and L. Goubin, "On Boolean and Arithmetic Masking Against Differential Power Analysis," in Ç. Koç and C. Paar, (eds.), *Cryptographic Hardware and Embedded Systems—CHES 2000. Proceedings of the Second International Workshop*, Vol. 1965, Worcester, MA: Springer-Verlag, August 2000, pp. 231–237.

[81] Shamir, A., "Protecting Smart Cards from Passive Power Analysis with Detached Power Supplies," in Ç. Koç and C. Paar, (eds.), *Cryptographic Hardware and Embedded Systems—CHES 2000. Proceedings of the Second International Workshop*, Vol. 1965 of *Lecture Notes in Computer Science*, Worcester, MA: Springer-Verlag, August 2000, pp. 71–77.

CHAPTER

10

Contents

Conclusion

Complete content protection by technical means is, as has been discussed in the preceding chapters, a goal that may well remain elusive for the application area where it is has been most sought after in recent decades, namely, in the prevention of unauthorized duplication and distribution of creations such as software, music, and movies by consumers and other pirates on the part of the content industry.

10.1 Digital rights management systems

This dichotomy between the desirable and the feasible is most evident in the case of digital rights management systems for consumer-oriented multimedia data. The application requirements for inexpensive implementability and retaining backwards compatibility, even with systems that are known to be compromised, sharply limit the effort that can be expended on securing the management functionality itself. For devices or software that are placed under the physical control of end users, protection against tampering against adversaries even with modest skills becomes—as discussed in Chapters 8 and 9—quite difficult given these boundary conditions.

The knowledge of circumvention or, as in the case of satellite or cable pay TV, even specialized hardware for this purpose, is very likely to be disseminated uncontrollably worldwide within a short period of time. The prospects for stopping such publication, which would have to be exhaustive to have any effect at all, are dim at best even if significant efforts are expended for the purpose of creating a precedent, and cost-effectiveness in an individual case is not a concern. As a result, DRM systems that require considerable investments for licensing or in-house development even apart from increases in costs per unit have historically been effective for weeks or months at most. This calls the overall effectiveness of such technical measures into question, since not only do the costs for developing or licensing and implementing the DRM mechanisms have to be amortized against lost sales, but the same is also true for costs incurred after the protection mechanism has been broken (e.g., increased production cost or per-unit licensing). As noted in Chapters 1 and 8, in at least partial recognition of this underlying conundrum for rights owners, the major associations of rights owners from the music and film industry have actively pursued legal protection not only in a unified copyright schema for the signatory nations of the WIPO treaties, but also in the form of the express prohibition of the circumvention of technical means of protection [1].

This raises a number of questions, not the least of which is why an individual or organization that is willing to violate copyright law would be deterred by another law that threatens not the result of piracy but rather a specific form of it.

Conversely, if the condition for prosecution lies in the circumvention of a protection mechanism, for which—as discussed throughout this book—it must be assumed that any mechanism satisfying the bounding requirements for cost and convenience can and will ultimately be broken assuming that a sufficient demand exists for the data or functionality protected, then the actual quality of a protection system would indeed be irrelevant, as the mere existence of such a mechanism would be sufficient. Viewed this way, the problem of content protection is again reduced to a legal one—and this implies that the effectiveness would have to be largely based on the expectation of discovery and prosecution of potential violators.

10.1.1 Privacy, competition, and freedom of speech

As discussed in Chapter 1, modern copyright law has its origins in censorship; the restriction of the right to print books to a limited number of easily controlled entities permitted the effective enforcement.

A truly effective DRM has at least the potential for similar restrictions if, as in the case of the statute of Queen Mary, it is bolstered by legislation that prohibits the circumvention of DRM mechanisms or even mandate their use.

One of the most immediate threats, however, is not that of censorship but rather the use of DRM mechanisms to limit competition and create closed markets with a captive customer base. If a particular DRM is successful for a given market segment (i.e., if a critical mass of attractive content is reached) and requires the purchase of specific software or even hardware, then the barrier for competitors rises considerably, since customers need to be convinced to invest for a second or third device and would also be inconvenienced by switching between mechanisms. Hence, even if no monopoly or oligopoly that rises to the level of antitrust regulations is in fact created, the DRM whose ostensible purpose is ensuring that copyright law is enforced may well act against the objective behind modern copyright law in restricting the selection and access to creations [2].

Privacy, although a somewhat amorphous concept that is treated very differently in various jurisdictions [e.g., primarily in the context of tort law in the United States [3, 4] or as part of fundamental constitutional rights in Germany (1 BvR 209/83; NJW 84, 419)], can be said to include at least the expectation that an individual has the right to fend off intrusion into private affairs and the public disclosure of potentially embarrassing private information [5]. Information of the pervasive and all-encompassing type potentially gathered and processed by DRM systems, which may include information such as reading or listening habits of a very personal nature, is very likely to fall under this heading regardless of the jurisdiction of origin. This has the potential of placing the customer a priori into an adversarial relation with the DRM operator. Furthermore, as argued by Samarajiva in [6], the trust of consumers in an organization or business engaged in collecting information on transactions, usage habits, and other profiling of individuals in a coercive (even if only in that a product or service provided is not available unless consent to the collection of information is given) or clandestine fashion will likely weaken over time. This can cause consumers to resist the collection or to deliberately provide misleading information, which in turn causes the collecting entities to pursue the information all the more aggressively, leading to a vicious cycle that is ultimately counterproductive for the collecting entities.

The paragraphs above can only barely touch upon the important privacy issues immanent in content protection systems; for further details refer to [6–9].

Placing issues of freedom of speech—one of the most cherished freedoms throughout the civilized world—in the context of pervasive digital rights management may seem exaggerated, but one needs only to consider the potentially chilling effect that ubiquitous surveillance of actions by end users can have if a less than perfectly benign environment and governmental (or even employer) benevolence is stipulated. If individuals refrain from recording thoughts on digital media or from communicating these to others because of an expectation that this speech will not only be recorded and attributed

(i.e., the absence of a reasonable expectation of anonymity) but also correlated to other actions that can be presented in a possibly damaging way, even
if the individual actions themselves and the totality of the data collected are
completely benign, then the freedom of speech is in fact impeded.

Besides this immediate potential effect of pervasive DRM deployment, the
same potential limitation of speech can also be observed in the concomitant
legislation enforcing the WIPO/WCT treaties such as *Universal City Studios,
Inc. v. Corley* (2001) or *Felten, Liu, Craver, Wu, Wallach, Swartzlander, Stubble-
field, Dean, and USENIX Association v. Recording Industry Association of America,
Inc., Secure Digital Music Initiative Foundation, Verance Corporation, John Ashcroft
in his official capacity as attorney general of the United States* (2001) [10]. While
in most of the legislation implementing the WIPO/WCT treaties there exists
an exemption for academic research in security mechanisms and protocols
including circumvention of protective measures, the risk not only applies to
individual researchers that may well be willing to accept it but also creates a
situation where the institutions employing and supporting such researchers
are exposed to litigation and may be unwilling to do so in the future. Such
restrictions are not only damaging for research but also for the general public,
since research into vulnerabilities and defects of computer systems and software beyond DRM may well be stifled to the detriment of security and reliability of computer systems overall as economic incentives for producing
defect-free systems decline even further than is already the case due to general abrogations of all warranties and liabilities [11–13].

10.1.2 DRM in commercial environments

The situation described above becomes markedly less bleak if one no longer
has to stipulate that a DRM system is subject to tampering both arbitrary
in duration and level of sophistication and, moreover, that an attacker can
obtain multiple instances of the DRM protection mechanism for progressive
efforts at tampering in case a system under attack is either destroyed during
analysis and tampering or is able to successfully engage tamper-responding
mechanisms.

This, however, is typically the case for (multimedia) data handled in commercial environments. Not only are such environments typically administered centrally so that an individual does not enjoy the level of privileges in
operating general purpose computers that enable tampering with software-
based DRM systems in the consumer environment, but given modest precau-
tions such as alarm systems reacting to the opening or similar manipulation

of device enclosures, most attempts at physical tampering are unlikely to progress far without the physical security mechanisms and staff of the commercial entity in question being alerted and removing the potential tamperer.

The parameters for such application scenarios influencing the absolute efficacy of a DRM also differ significantly from the scenario discussed in the preceding section. The value of data in terms of research and development expenditures, opportunity values, and even for protection against litigation can justify much more sophisticated—and therefore expensive and also intrusive—protective mechanisms. This, however, is not purely limited to traditional access control; integrated protection can also include most of the multimedia formats discussed in this book, such as design drawings from research and engineering divisions, video material such as automobile industry tests and telemetry including potentially sensitive information on the visual appearance and safety characteristics of early prototypes, and three-dimensional geometric models used not only in traditional mechanical engineering and research but also increasingly representing considerable values in the chemical and pharmaceutical industry where such models are used to represent chemical compounds and their reactions ranging to complex protein interactions that elucidate the behavior of advanced medications.

With the threat of physical tampering reduced and privileged access to software environments curtailed (and under the assumption that personnel with elevated privileges are not compromised or act outside the law), a DRM system must still contend with other potential vulnerabilities even assuming that the mechanisms of the DRM itself are in fact free of defects. The most prominent threat to such a mechanism implemented on a general-purpose computer is constituted by the class of privilege elevation attacks on the underlying operating system. A Trojan horse or user that is able to obtain such privileges is free to disable the DRM mechanism or conduct sophisticated tampering attacks against the DRM system. It is therefore a concomitant requisite of any DRM-enabled system that the correctness and robustness of the underlying hardware and software is at least equal to the level of protection afforded by the DRM itself.

Another threat that gains particular relevance for DRM in commercial settings are the potential for denial-of-service attacks or losses of access and availability of protected data due to attacks, faults, or mishandling in the DRM itself. Given the value of such commercial data as well as other considerations such as legal requirements for long-term archiving of certain data compared to the mere nuisance in the case of a similar malfunction in a consumer environment, great care must be exerted in both the creation of the mechanism itself and in the design of disaster recovery mechanisms. As with most other security systems, the conflict between protection against disasters and other failures and the need to operate a seamless protection system is also evident in the case of DRM systems. As an example, while it may be desirable to retain archival copies of unencrypted data for the eventuality of catastrophic failure on the part of the DRM (e.g., loss of all decryption keys—such an event can occur through failures or as part of a deliberate denial-of-service attack that triggers tamper response functionality in the desired locations), the only barrier between an adversary and the data is no longer the correctness and completeness of the DRM but rather the physical security of the backup data repository.

10.2 Digital watermarking

The problems alluded to in Section 10.1, however, justify the development and deployment of mechanisms that permit ex post facto identification of unauthorized copies or, depending on the application scenario, even specific traitor tracing mechanisms. Given their ability to survive multiple conversions between digital and analog representations, independence of encoding, and the ability to survive at least some manipulations and attacks, digital watermarks currently represent the most promising approach for this purpose.

However, even assuming sufficient robustness of a given digital watermark, the precise location of embedding represents a possible vulnerability locus. If the marking is embedded or retrieved in consumer electronics or general-purpose computers under the control of end users, then the embedding and retrieval mechanism is potentially vulnerable to tampering under the assumptions described in Section 9.2. Unless an application scenario can minimize the risk of tampering to an acceptable level, the embedding of the digital watermarks identifying the customer or carrying a unique serial number would preferably have to occur outside the domain of control of potential adversaries. This protection, however, would have to be balanced against the increased cost of personalization in the case of physical media.

Another aspect that strongly suggests the use of digital watermarks as a protection mechanism is the observation described in Chapter 1 that many listeners (this phenomenon is largely limited to the audio domain thus far) find analog copies, particularly first generation copies that are possibly created using digital recording devices, perfectly acceptable in terms of the quality offered. Any protection afforded by a DRM system by definition ends at the point where signals are converted to analog representation, so an attacker must simply tap into this signal path. It is rather unclear whether such an approach would constitute "circumvention" under even the most stringent interpretation of the requirements set forth in the WIPO/WCT treaties.

10.3 Outlook

The various technical means for content protection, some of which we were able to briefly present in the present volume, represent a challenging and active area for research and development and has, particularly in the case of digital watermarks for various multimedia data types from still images to three-dimensional geometry models, witnessed a veritable explosion of results over the past decade.

Nevertheless, for digital watermarks to be useful in content protection application scenarios, a significant amount of research is still required particularly in the area of increasing robustness to the various types of attacks described in Chapter 7.

A number of the attacks described in Chapter 7 can be classified as protocol-specific or usage-specific and hence must be dealt with by way of careful design of protection mechanisms, although unfortunately some of the most desirable deployment scenarios (where complete detectors or even embedders and possibly key material are present in the hands of adversaries) are vulnerable to classes of attacks such as oracle attacks.

There exists a need for additional research in digital watermarking particularly with regard to improved robustness against the various types of desynchronization attacks. Such robustness can presumably be achieved if more information (i.e., signal-specific feature level semantics) can be used for registration of signals that have been subjected to manipulation or outright deliberate desynchronization attacks. Similar improvements can also be conjectured for signal processing attacks if higher level semantics are considered based on the significant improvements in robustness achieved in the past with the introduction of advanced perceptual modeling into digital watermarking.

Moreover, for some application scenarios discussed here, such as monitoring applications, any such improved signal registration and detection algorithm must be limited in its computational complexity so that the detection of a watermark (and hence of a license or other copyright violation) can be accomplished in real time and with commensurate expenses for equipment in the case of audio and video data.

The constraint of computational complexity is less severe in the case of off-line detection and verification, but, particularly in the case of video data, the signal bandwidth required for processing imposes a practical limit on complexity.

Any use of protection mechanisms—and indeed multiple redundant or integrated mechanisms are highly desirable—must be balanced carefully between multiple conflicting requirements such as the initial and per-unit cost of the mechanism, the extent to which customers may be inconvenienced and thus kept from purchasing either the product or subsequent offerings, and the actual protection.

Many of the restrictions on the efficacy and effectiveness of technical content protection mechanisms discussed in this volume are, moreover, not primarily caused by purely technical reasons, but by pragmatic considerations. Future developments in media and distribution such as breakthroughs in on-line presentation of multimedia content over more traditional off-line, storage-based approaches may well shift the balance compared to the discussions here [2]. If anything, however, the protection of intellectual property in an increasingly digital domain with abundant computational capabilities and bandwidth capacity will remain a multidimensional challenge with technical, economical, and legal aspects demanding equally careful attention.

References

[1] Marks, D. S., and B. H. Turnbull, Workshop on Implementation Issues of the WIPO Copyright Treaty (WCT) and the WIPO Performances and Phonograms Treaty (WPPT), *Technical Protection Measures: The Intersection of Technology, Law and Commercial Licenses*, World Intellectual Property Organization, December 1999.

[2] Rifkin, J., *The Age of Access*, New York: Penguin Putnam, 2000.

[3] Warren, S., and L. Brandeis, "The Right to Privacy," *Harvard Law Review*, Vol. 4 No. 5, December 1890, pp. 193–220.

[4] Alderman, E., and C. Kennedy, *The Right to Privacy*, New York: Vintage, 1997.

[5] Prosser, W. L., "Privacy, A Legal Analysis," *California Law Review*, Vol. 48, 1960, pp. 338–423.

[6] Agre, P., and M. Rotenberg, *Technology and Privacy: The New Landscape*, Cambridge, MA: MIT Press, 1997.

[7] Westin, A. F., *Privacy and Freedom*, Gateshead, U.K.: Athenaeum, 1967.

[8] Diffie, W., and S. Landau, *Privacy on the Line*, Cambridge, MA: MIT Press, 1998.

[9] Garfinkel, S., *Database Nation*, Sebastopol, CA: O'Reilly & Associates, 2000.

[10] Craver, S. A., et al., "Reading Between the Lines: Lessons from the SDMI Challenge," in *Proceedings of the 10th USENIX Security Symposium*, Washington D.C., August 2001.

[11] Needham, R. M., and R. J. Anderson, "Programming Satan's Computer," in J. van Leeuwen, (ed.), *Computer Science Today: Recent Trends and Developments*, Vol. 1000 of *Lecture Notes in Computer Science*, Heidelberg: Springer-Verlag, 1995, pp. 426–440.

[12] Anderson, R., "Security in Open Versus Closed Systems—The Dance of Boltzmann, Coase and Moore," *Proceedings of Open Source Software: Economics, Law and Policy*, Toulouse, France, June 2002.

[13] Varian, H. R., "System Reliability and Free Riding," *Proceedings of Open Source Software: Economics, Law and Policy*, Toulouse, France, June 2002.

Glossary

k-means clustering   Clustering of data is a method by which large sets of data are clustered into groups of smaller sets of similar data. A clustering algorithm attempts to find natural groups of components (or data) based on some similarity. *k*-means clustering minimizes the average error of the data samples to its corresponding cluster center.

Bark   A non-SI unit of measurement named after the physicist Barkhausen, modeling the representation of sound in the human auditory system more closely than SI units. The Bark scale assumes that one unit corresponds to a constant length (1.3 mm) along the basilar membrane. The Bark frequency scale ranges from 1 to 24 consisting of center frequencies and band edges to be interpreted as samplings of a continuous variation in the frequency response of the ear to a sinusoid or narrowband noise signal, corresponding to the first 24 critical bands of hearing and range up to 15.5 kHz, implying that the highest sampling rate for which the Bark scale is defined up to the Nyquist limit is 31 kHz.

codec   1.   In communications engineering, the term *codec* is an acronym for "encoder/decoder" and is used in reference to integrated circuits performing data conversion.
2.   The term *codec* is also an acronym that stands for "compression/decompression."

Cryptography   From the Greek χρυπτός (hidden, secret) and γραφειν (to write), literally "secret writing."

EBU   European Broadcasting Union.

Entropy coding   Entropy encoding exploits the nonuniformity of the probability distribution of the input symbols for an efficient bit allocation for each symbol.

255

IEEE1394   IEEE Standard 1394, a high-speed serial bus protocol and device specification family which includes specifications for isochronous and asynchronous data transfer and integrates protection mechanisms for interdevice and intradevice communication.

Manifolds   A homomorphic mapping from a topological space $\mathcal{A}$ to a topological space $\mathcal{B}$ is a continuous mapping whose inverse is also continuous. Two topological spaces $\mathcal{A}$ and $\mathcal{B}$ are homomorphic if there is a homomorphic mapping from $\mathcal{A}$ to $\mathcal{B}$. A topological space $\mathcal{S}$ is an n-manifold if for every point $\mathbf{p}$ in $\mathcal{S}$, there is an open set $\mathcal{U}_p$ containing $\mathbf{p}$ in $\mathcal{S}$ such that $\mathcal{U}_p$ is homomorphic to the n-dimensional open sphere in the n-dimensional Euclidean space.

MESH   A software for measuring error between surfaces using the Hausdorff distance (developed at EPFL, Lausanne).

MITI   Ministry of International Trade and Industry, since reorganized into the METI, or Ministry of Economy, Trade and Industry (Japanese: Kezai Sangyo Sho).

PCI   The peripheral component interconnect is an intradevice interconnection protocol and component specification family. There exist a number of variations of PCI, including CompactPCI, Mini PCI, Low-Profile PCI, Concurrent PCI, and PCI-X.

Quantization   In quantization, the input values are replaced by a fixed subset of representative values.

SDMI   Secure Digital Music Initiative. The SDMI consists of a consortium of music companies trying to standardize technologies for the protection of digital music with regard to playing, storing, and distributing.

Steganography   From the Greek $\sigma\tau\varepsilon\gamma\breve{\alpha}\nu\acute{o}\varsigma$ (closely covered, sheathed) and $\gamma\rho\acute{\alpha}\varphi\varepsilon\iota\nu$ (to write), literally "hidden writing."

USB   The Universal Serial Bus family of protocols provides a specification for protocols and devices interconnected via a serial bus for interdevice communication.

List of acronyms

AAC	advanced audio coding
ACR	absolute category rating
A/D	analog to digital
AEAD	average edges angles difference
AGC	automatic gain control
ANOVA	Analysis of variance
AVO	audio visual object
BPM	blind pattern matching
CAD	computer-aided design
CCI	copy control information
CD	compact disc
CD-R	compact disc recordable
CD-ROM	CD read-only memory
CGMS	Copy Generation Management System
COTS	commercial off the shelf
CPU	central processing unit
CQS	continuous quality scale
CRL	certificate revocation list
CSF	contrast sensitivity function
CSG	constructive solid geometry
CSS	content scrambling system
CT	computed tomography

D/A	digital to analog
DAM	dot-area modulation
DAT	digital audio tape
DCC	digital compact cassette
DCT	discrete cosine transform
DFT	discrete Fourier transform
DHBP	data hiding in block parity
DICOM	digital imaging and communications in medicine
DMCA	Digital Millennium Copyright Act
DPM	dot-position modulation
DRM	digital rights management
DSP	digital signal processor
DTCP	Digital Transmission Content Protection Specification
DTLA	Digital Transmission License Administrator
DVD	digital versatile disk
DVD-CCA	DVD Copy Control Association
DVI	digital video interface
DWT	discrete wavelet transform
EBU	European Broadcasting Union
EC-DSA	Elliptic Curve Digital Signature Algorithm
EPFL	Ecole Polytechnique Fédérale de Lausanne
ESDR	edges standard deviation ratio
FEM	finite element method
FFD	free-form deformation
FFT	fast Fourier transformation
FIR	finite impulse response
GIS	geographical information systems
GPU	graphical processor unit
HDCP	High-Bandwidth Digital Content Protection protocol
HVS	human visual system

ICE	in-circuit emulator
ICM	iterated conditional modes
IEEE	Institute of Electrical & Electronics Engineers
IFPI	International Federation of the Phonographic Industry
IFS	iterated function systems
ITU	International Telecommunication Union
ITU-R	Radiocommunication Sector of the ITU (former CCIR)
IVK	integrity verification kernel
JAWS	Just Another Watermarking System
JND	just noticeable difference
JPEG	Joint Picture Expert Group
LFSR	linear feedback shift register
LPM	log-polar mapping
LSB	least significant bit
MCLT	modulated complex lapped transform
MDP	mesh density pattern
MEP	macroembedding primitive
MF	matched filtering
MIDI	Musical Instrument Digital Interface
MPEG	Motion Picture Expert Group
MRI	magnetic resonance imaging
MSE	mean-square error
MUSHRA	multistimulus with hidden reference anchors
NBE	Normal Bin Encoding algorithm
NTSC	National Television Standards Committee
NURBS	nonuniform rational B-spline
OC	operation characteristic
OCR	optical character recognition
ODG	objective difference grade
OMR	optical music recognition

PAL	permissive action link
PAL	phase alternation line
PAM	psychoacoustic model
PCA	principal component analyses, also known as the Karhunen-Loeve transformation
PCI	Peripheral Component Interconnect protocol
PCM	pulse code modulation
PDM	perceptual distortion metric
PEAQ	perceived audio quality
PN	pseudonoise
PRNG	pseudorandom number generator
PSC	power-spectrum condition
PSNR	peak signal-to-noise ratio
QIM	quantization index modulation
RMS	root mean square
RMSE	root-mean-square error
ROC	receiver operating characteristics
ROI	region of interest
RST	rotation, scaling, and translation
SCMS	Serial Copy Management System
SDG	subjective difference grade
SDMI	Secure Digital Music Initiative
SECAM	Systéme electronique couleur avec memoire
SME	small and medium enterprises
SMR	signal-to-mask ratio
SNHC	synthetic and natural hybrid coding
SNR	signal-to-noise ratio
SPL	sound pressure level
SR	stored reference
SRM	System Renewability Message
SRMV	SRM, 16-bit version number

SSP	self-spanning patterns
TCPA	Trusted Computing Platform Alliance
TFA	triangle flood algorithm
TPA	trusted platform agent
TPM	trusted platform module
TR	transmitted reference
TSPS	Triangle Strip Peeling Symbol Sequence
TSQ	triangle similarity quadruple
TVR	tetrahedral volume ratio
USB	Universal Serial Bus
VBI	vertical blanking interval
VFA	Vertex Flooding Algorithm
VQEG	Video Quality Experts Group
WIPO	World Intellectual Property Organization
WMS	watermarking minimum segment
WWW	World Wide Web

About the authors

Michael Arnold studied physics at the Julius-Maximilians University of Würzburg, where he received his Dipl.-Phys. in December 1994. Since February 1996, he has been working as a researcher at the Fraunhofer Institute for Computer Graphics. Mr. Arnold is currently the head of the watermarking group in the security technology department at the Fraunhofer Institute for Computer Graphics in Darmstadt, Germany.

Martin Schmucker has been working with the Fraunhofer Institute for Computer Graphics in Darmstadt, Germany, in the security technology department, since 2000. Before this, he worked in industry in the field of telematics and traffic control systems. His research interests include content protection and digital watermarking.

Stephen D. Wolthusen is currently the deputy department head of the security technology department at the Fraunhofer Institute for Computer Graphics in Darmstadt, Germany, and has been involved in computer security and information assurance research since 1993.

Index

A

Recent Titles in the Artech House Computer Security Series

Rolf Oppliger, Series Editor

For further information on these and other Artech House titles,
including previously considered out-of-print books now available through our
In-Print-Forever® (IPF®) program, contact:

Artech House
685 Canton Street
Norwood, MA 02062
Phone: 781-769-9750
Fax: 781-769-6334
e-mail: artech@artechhouse.com

Artech House
46 Gillingham Street
London SW1V 1AH UK
Phone: +44 (0)20 7596-8750
Fax: +44 (0)20 7630-0166
e-mail: artech-uk@artechhouse.com

Find us on the World Wide Web at:
www.artechhouse.com

Recent Titles in the Artech House Computing Library

Advanced ANSI SQL Data Modeling and Structure Processing, Michael M. David

Advanced Database Technology and Design, Mario Piattini and Oscar Díaz, editors

Action Focused Assessment for Software Process Improvement, Tim Kasse

Building Reliable Component-Based Software Systems, Ivica Crnkovic and Magnus Larsson, editors

Business Process Implementation for IT Professionals and Managers, Robert B. Walford

Data Modeling and Design for Today's Architectures, Angelo Bobak

Developing Secure Distributed Systems with CORBA, Ulrich Lang and Rudolf Schreiner

Future Codes: Essays in Advanced Computer Technology and the Law, Curtis E. A. Karnow

Global Distributed Applications with Windows® DNA, Enrique Madrona

A Guide to Software Configuration Management, Alexis Leon

Guide to Standards and Specifications for Designing Web Software, Stan Magee and Leonard L. Tripp

Implementing and Integrating Product Data Management and Software Configuration, Ivica Crnkovic, Ulf Asklund, and Annita Persson Dahlqvist

Internet Commerce Development, Craig Standing

Knowledge Management Strategy and Technology, Richard F. Bellaver and John M. Lusa, editors

Managing Computer Networks: A Case-Based Reasoning Approach, Lundy Lewis

For further information on these and other Artech House titles, including previously considered out-of-print books now available through our In-Print-Forever® (IPF®) program, contact:

Artech House
685 Canton Street
Norwood, MA 02062
Phone: 781-769-9750
Fax: 781-769-6334
e-mail: artech@artechhouse.com

Artech House
46 Gillingham Street
London SW1V 1AH UK
Phone: +44 (0)20 7596-8750
Fax: +44 (0)20 7630-0166
e-mail: artech-uk@artechhouse.com

Find us on the World Wide Web at:
www.artechhouse.com